T0180608

Springer Series in Statistics

Springer Series in Statistics

(continued after index)

Marco A.R. Ferreira
Herbert K.H. Lee

Multiscale Modeling

A Bayesian Perspective

 Springer

Marco A.R Ferreira
Department of Statistics
146 Middlebush Hall
University of Missouri
Columbia, MO 65211
USA
ferreiram@missouri.edu

Herbert K.H. Lee
School of Engineering
Department of Applied Mathematics
  and Statistics
University of California
Santa Cruz, CA 95064
USA
herbie@ams.ucsc.edu

ISBN 978-1-4419-2426-1   e-ISBN 978-0-387-70898-0

Printed on acid-free paper.

springer.com

To Alejandra

Preface

A wide variety of natural processes occur on multiple scales, either naturally or as a consequence of measurement. This book contains methodology for the analysis of data that arise from such multiscale processes. This topic is relatively young in statistics, and our goal in writing this book is to bring together a number of recent developments and make them accessible to a wider audience. Here we focus on three main classes of methods: multiscale random field models, approximate basis representations, and implicit methods. Multiscale random field models attempt to directly model the process at each scale as well as provide a model for the link between scales; they build upon standard autocorrelated time series and intrinsic spatial processes to explicitly model multiscale correlated processes. Approximate basis representations include wavelets and convolution methods, both of which involve a finite approximation to an infinite basis set for the space of continuous functions, where finer scales use successively more basis elements. Implicit methods induce a multiscale structure via computation, having the various scales share information as the model is being fit but not directly modeling the relationship between scales.

We take a Bayesian approach, which allows for full accounting of uncertainty. The methodology presented herein allows one to deal with the delicate issue of uncertainty at multiple scales. The Bayesian approach also facilitates the use of knowledge from prior experience or data, and these methods can handle different amounts of prior knowledge at different scales, as often occurs in practice.

We provide a number of real-world examples that are thoroughly analyzed in order to demonstrate our methods and to assist readers in applying these methods to their own work. While this is a high-level book, our primary objective is to make the material accessible so that the reader can implement the methods and hopefully find them helpful in practice. To further assist readers, we are making source code (for R) available for many of the basic methods discussed in this book. Code can be found at http://www.ams.ucsc.edu/~herbie/multiscale.

The book is aimed at statisticians, applied mathematicians, and engineers working on problems dealing with multiscale processes in time and/or space, such as in petroleum engineering, materials engineering, signal processing, finance, environmetrics, and geology. This book will also be of interest to researchers in academia working on multiscale computational problems, and can be used as the basis for an upper-level graduate seminar.

The main prerequisite for this book is a knowledge of Bayesian statistics at the level of Gelman et al. (1995) or a similar book. Such a background should necessarily include familiarity with basic Markov chain Monte Carlo methods. Knowledge of time series at the level of Peña et al. (2000) is helpful to understand Chapters 3 and 12.

A significant portion of the methodology highlighted in this book originated from work done as part of the National Science Foundation funded (DMS 9873275) multidisciplinary project "Multiscale Modeling and Simulation in Scientific Inference: Hierarchical Methods for Parameter Estimation in Porous Flow", which brought together statisticians, applied mathematicians, engineers, and geologists from several universities to study multiscale processes in geology and hydrology. Both authors were members of the Center for Multiscale Modeling and Distributed Computing at Duke University during that time, where they developed several of the methods included in this book. Marco Ferreira also received partial support from CNPq Brazil grants 302717/2003-0 and 402010/2003-5. Herbert Lee also received partial support from National Science Foundation grants DMS 0233710 and DMS 0504851.

There are many people who have contributed in one way or another to the writing of this book. We would like to thank the editor, John Kimmel, for his incentive, and the former editor, Stephanie Harding, for convincing us to write this book. We have benefitted from discussions about Bayesian statistics and/or multiscale modeling with several colleagues, among them Mike West, David Higdon, Chris Holloman, Peter Müller, Brani Vidakovic, Dani Gamerman, Alexandra Schmidt, Hedibert Lopes, and Helio Migon. A significant portion of the book was written while Marco Ferreira was a faculty member at the Institute of Mathematics, Federal University of Rio de Janeiro. The authors would like to thank Robert Gramacy for his extensive help with the R source code; Adelmo Bertolde, Geraldo Cunha, Luzia Tonon, and Vinicius Israel for the implementation of the examples presented in Sections 8.5, 9.2.4, and 9.3.4; and Albert Ko for giving permission to use the leptospirosis data in the application presented in Section 9.3.4.

Columbia, Missouri, USA   *Marco A.R. Ferreira*
Santa Cruz, California, USA   *Herbert K.H. Lee*
April 2007

Contents

Part IV Implicit Multiscale Models

Part I

Introduction

1

Introduction

In this book, we are concerned with multiscale methods and models. The term multiscale broadly refers to processes, algorithms, and data that can be structured by scale. A well-known example of a multiscale stochastic process is a fractal (e.g., Mandelbrot, 1999). An example of a multiscale algorithm is the fast wavelet transform (e.g., Vidakovic, 1999; Mallat, 1999). An example of multiscale data is a time series dataset that happens to be observable at different sampling frequencies. One could check the price of a stock once a year, daily, hourly, or every minute. The fluctuations in the annual prices will generally have rather different behavior than the hourly fluctuations, and one may want to model both sorts of behavior simultaneously.

The goal of this book is to present methodology for dealing with multiscale situations. These methods represent a number of rather different perspectives on the multiscale problem. What they have in common is their value in multiscale modeling. In all cases, these methods are based on probability models. Furthermore, we take a fully Bayesian perspective, and the methods contained herein are all amenable to a Bayesian approach. There are many more multiscale methods in the literature, and we are not able to discuss all of them in a single book. We focus on those that are the most suitable for Bayesian models. The Bayesian paradigm allows full accounting for uncertainty, something that can be quite difficult in complex models, such as multiscale models. The problem of propagating uncertainty between scales makes the Bayesian approach particularly attractive.

We note that the terms *multiscale* and *multiresolution* are being used increasingly in the literature in a growing variety of disciplines and applications. We will not attempt to distinguish between these two terms but will essentially use them interchangeably.

When does it pay to use multiscale ideas? Throughout this book, we focus on three main cases where the multiscale approach is particularly helpful: for processes that are naturally multiscale, when information is available at different levels of resolution, and when embedding a standard problem in a multiscale framework leads to significant computational advantages. Let us briefly consider each of these cases.

Processes may naturally occur in a multiscale fashion. In some cases, relevant features of the data can be seen only at particular scales, with different aspects requiring different scales. Effective modeling of such a process clearly requires a fully multiscale model. Another important case is when the process has a simpler representation at coarser scales. In that case, a multiscale model can be used to represent the process in a cascade manner from coarser to finer levels. But we also include in this categorization more general processes that have underlying multiscale mechanisms. An example is that of the concentration of a ground-level pollutant such as ozone. A station making hourly measurements over the course of a year or more will notice both a daily cycle (higher in midday) and an annual cycle (higher in the summer) in addition to the possible fine-level deviations at the level of measurement. Thus a model would need to incorporate both annual and daily components. The multiscale approach provides a natural mechanism for dealing with this situation. Note that data may only be available at a single level of resolution. Typically the data will be at the finest level of resolution being considered, but some methods such as those in Chapter 10 can also deal with the case where the data occur on a coarser scale than the finest scale of interest.

The other obvious application for multiscale methods is when data are collected at multiple levels of resolution. The time series example given at the beginning of this chapter is an example of this, where the time series may be observed at different levels of resolution, possibly by different observers, and the two sets of data need to be combined. Another example is in geology, where soil characteristics may be measured using several different techniques, each resulting in data at a different physical scale of interpretation. Core samples provide information at a point but no direct information away from the sampling location. Flow experiments can provide information on an intermediate scale, covering more area but not providing precise information at specific locations. Seismic experiments are at a yet larger scale, providing information about larger areas. To fully model the ground, it may be necessary to incorporate all three sources of information in a statistically coherent manner. This example will be revisited in more detail in Chapter 16. In these multiscale data cases, we often consider the true process to occur at a particular scale, and it is just the data that are collected at multiple scales that drive the need for a multiscale approach. Of course, it is possible to have both the data and the underlying process be multiscale, which would be a combination of both this case and that of the previous paragraph.

A third set of problems amenable to multiscale approaches are ones where multiple levels of resolution are artificially imposed in order to improve the tractability of the problem by helping with computational aspects. Modern Bayesian computation is often done with Markov chain Monte Carlo (MCMC) methods, and in complex problems the chain may mix slowly, requiring unreasonable amounts of time to converge. In addition, there may be a large number of posterior local modes, and the chance of the chain becoming trapped around one of these modes is very high. By creating an artificial coarser scale, the parameter space can be reduced, resulting in many fewer local modes and allowing the model to be fit more easily on the coarser scale, thus significantly improving mixing. The coarser scale is then related back to the original finer scale. One example of this occurs in medical imaging, such as with single photon emission computed tomography (SPECT). The problem of interest is to reconstruct properties of the object being scanned from the collected photons. This information is desired on a particular scale, and a physical model relates these parameters to the observed counts. Running this model can be computationally expensive, and the correlations in the parameters (the object properties on a grid) can cause poor mixing of MCMC methods. Fitting the model at a coarser resolution (i.e., a coarser grid for the unknown object properties) reduces both the physical model run time and the number of unknown parameters, thus helping reduce the amount of time needed to use MCMC methods effectively. Note that the process is only of interest at a single resolution, the data are only collected at one resolution, and the multiscale model is used solely as a computational tool. The SPECT example is explored more fully in Chapter 17.

This book is divided into five parts. This introductory chapter plus chapters on basic spatial models and our primary illustrative example comprise Part I. Part II concerns multiscale decomposition methods, in particular convolutions and wavelets. Part III presents explicit multiscale models, including multiscale models on trees, multiscale random fields, multiscale time series, and change of support models. Part IV presents multiscale models that are implicitly defined by a computational linkage of the different levels of resolution; such linkages include Metropolis-coupled methods and genetic algorithms. Part V closes the book with some case studies. Each of the three main parts starts out with an overview chapter introducing the methods and putting them into broader context.

Finally, we note that basic computer code (in R) is available for many of the methods discussed in this book, and can be found at

http://www.ams.ucsc.edu/~herbie/multiscale

This code is not meant to be a complete ready-to-use package but rather a starting point from which the user may make modifications for their particular situation. The routines illustrate the key aspects of the methods for basic cases and can easily be adapted to more complex problems.

2

Models for Spatial Data

Before launching into the main topics of the book, we first want to introduce two standard models used for spatial data, as they will reappear throughout the book. The first is the Markov random field (MRF), which is most useful for grids and irregular areal data. The second is the Gaussian process, which is more useful when a continuous surface is desired or a wider variety of spatial smoothness needs to be specified or fit.

2.1 Markov Random Fields

Markov random fields are typically defined over a regular lattice but can also be used for irregular grids or regular or irregular areal units. Here we focus on implementation on a regular grid, as that is the most relevant for the multiscale methods in the rest of the book. But we do want to stress that these models are flexible and useful well beyond the regular grid setting. Similarly, we consider only Gaussian MRFs but acknowledge that alternative forms of MRFs may be more applicable in other situations. Many more details of MRFs can be found in references such as Rue and Held (2005), Hjort and Omre (1994), and Besag (1974).

The key idea of a Markov random field is that the distribution of the process at a particular location depends only on the values of the process at neighboring locations. It is in this sense that it is Markovian—points outside the neighborhood are conditionally independent given the neighborhood. One of the degrees of flexibility in the model is the specification of the neighborhood. Often, only the immediately adjacent points will constitute the definition of the neighborhood. However, more extended neighborhoods are easily incorporated into the structure if desired. Here we typically focus on simple first-order neighborhoods (i.e., only the immediate neighbors). On a regular grid in two dimensions, this would be the four lattice points (or grid cells) that are horizontally or vertically adjacent. In three dimensions, there would be six such neighbors. In one dimension, there are only two neighbors. Of course, locations on the edges or corners of the lattice will have fewer neighbors. MRF models are closely related to conditional autoregressive (CAR) models (Banerjee et al., 2003).

Write the process values on the grid of spatial locations in vector form as $\mathbf{x} = (x_1, \ldots, x_n)$. A proper Gaussian Markov random field model, for example as considered by Ferreira and de Oliveira (2007), can be written as

$$\mathbf{x} \sim N(\mu \mathbf{1}_n, \Sigma), \tag{2.1}$$

where $\mu \in \mathbb{R}$ is a location parameter, $\mathbf{1}_n$ is an n-dimensional vector of ones, and $\Sigma^{-1} = \tau(\alpha I_n + H)$, with $\tau > 0$ a scale parameter, I_n the $n \times n$ identity matrix, and H is comprised of elements

$$H_{jk} = \begin{cases} h_k, & j = k, \\ -g_{jk}, & j \in N_k, \\ 0, & \text{otherwise}, \end{cases} \tag{2.2}$$

where $g_{jk} = g_{kj} > 0$ is a "measure of similarity" between sites j and k, N_k is the set of locations j such that j is a neighbor of k, and $h_k = \sum_{j \in N_k} g_{jk}$. In many cases, rotational symmetry of the spatial similarity will mean that g_{jk} will be the same for all (j, k) in the lattice. The parameter $\alpha > 0$ is a "spatial" parameter that controls the strength of association between the components of x and determines the main properties of model (2.1).

When $\alpha \to 0$, model (2.1) approaches the intrinsic autoregressive model (Besag et al., 1991; Besag and Kooperberg, 1995), which is an improper distribution that has been used extensively in spatial statistics as a prior distribution for latent processes or random effects (Sun et al., 1999; Carlin and Banerjee, 2003). As an example of the effects of the key parameters, Figure 2.1 shows perspective plots of random realizations of 2-D MRFs with mean zero and $g_{jk} = 1$ for all j and k. The left column shows intrinsic fields with $\alpha = 0$, and the right column shows proper (mean-reverting) fields with $\alpha = 1$. All plots are on the same scale. The rows show MRF scale parameters of $\tau = 1$, 4, and 16, and so they are successively smoother. Note that the proper fields are generally smoother than the intrinsic fields with the same scale parameter. The range of the intrinsic fields is typically larger because they are not mean-reverting, as the smoothness constraint is entirely locally defined.

In the intrinsic case, and assuming a zero-mean process, the conditional distribution for a point given its neighbors is

$$x_k | x_{-k} \sim N\left(\frac{-\sum_{j \neq k} H_{jk} x_j}{H_{kk}}, \frac{1}{\tau H_{kk}} \right), \tag{2.3}$$

where x_{-k} denotes all elements of $\mathbf{x}$ except x_k. For a first-order zero-mean process, this simplification is considerable. It also allows for efficiency in updating because the lattice can be partitioned into two checkerboards, as the neighborhood of each point is only the directly adjacent points, so picturing a

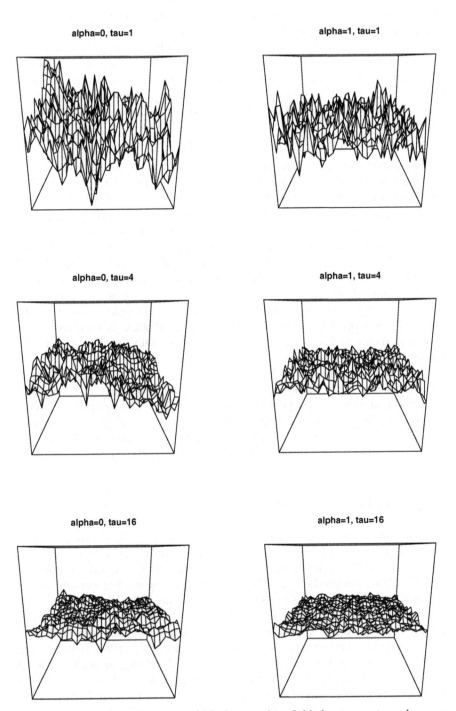

Fig. 2.1. Random realizations of Markov random fields by parameter values.

checkerboard, all of the black squares are conditionally independent given the values on the other half of the squares. Within a Bayesian context, fitting of the MRF is done via Gibbs sampling, such as with the complete conditional for each point given by Equation (2.3). Since the two colors of the checkerboard denote conditionally independent segments, the entire lattice can be updated in just two multivariate Gibbs steps. In practice, the MRF model **x** is combined with data **y**, such as with a conditionally independent normal likelihood

$$y_i|x_i \sim N(x_i, \sigma^2) \, ,$$

assuming that there is one observation for each cell (such as with image analysis) or using a suitable mapping of the MRF cells x_j to the data points y_i. This formulation also allows full updating with just two multivariate Gibbs steps. Conditionally conjugate priors of an inverse-gamma for σ^2 and a gamma for τ complete the specification, allowing Gibbs updates for those two parameters as well. By using all Gibbs steps, posterior inference is usually fairly efficient, with good mixing. In some more complex situations, such as the SPECT example of Chapter 17 with the computer model involved in the likelihood function, the MCMC updates require more care and achieving good mixing is more difficult (hence the value of a multiscale approach). Some ideas for other update proposals can be found in Lee et al. (2002). Alternative choices of priors, such as reference priors (Ferreira and de Oliveira, 2007), may also require Metropolis-Hastings updates.

From this development, it is clear how having a regular lattice makes the implementation of an MRF much simpler. For the irregular case, defining a neighborhood structure is possible, although it is a more complex process. In contrast, the Gaussian process models discussed in the next section have no additional complications for irregular grids because they are defined as a continuous process and can then be evaluated at any arbitrary set of locations.

2.2 Gaussian Processes

Gaussian processes have a long history of use for spatially distributed data. Much of the early work appeared in the geostatistical literature under the name of kriging (Matheron, 1963). Related ideas were explored initially as a method for interpolation, and later full probability models were developed (Journel and Huijbregts, 1978; Ripley, 1981; Cressie, 1993; Wackernagel, 1998; Stein, 1999). Applications can now be found in a wide variety of areas, including meteorological fields (Royle et al., 1999), unknown functions (O'Hagan, 1991), complex computer model responses (Sacks et al., 1989; Kennedy and O'Hagan, 2001; Santner et al., 2003; Fang et al., 2006), agricultural fertility gradients (Brownie et al., 1994), and pollutant fields (Host et al., 1995). Introductory descriptions of Gaussian processes can be found in Cressie (1993) and Hjort and Omre (1994). Gaussian processes are also used as a method of nonparametric regression to fit arbitrary surfaces in the presence of independent errors (Williams and Rasmussen, 1996; Neal, 1999). Lee (2004) gives more context for how Gaussian processes fit into the families of nonparametric regression methods. Detailed theoretical treatments are available in Abrahamsen (1997) and Stein (1999).

The key identifying feature of a Gaussian process is that it is a continuously defined process such that its values at any finite collection of locations have a multivariate Gaussian distribution. Two common simplifying assumptions are stationarity and isotropy. In its basic form, stationarity says that if you take any finite collection of points and a translation of that collection, they will both have the same distribution. Thus the process does not depend directly on location in the sense that the distributions of the points are the same everywhere. In particular, this implies that all points have the same mean and marginal variance. Considering for the moment a two-dimensional space, stationarity also implies that if you examine the joint distribution of a point and another point one unit away to the northwest, that bivariate normal distribution will be the same bivariate normal distribution for the joint distribution if you shift both points one unit to the east. This distributional invariance extends to any number of dimensions and to any finite number of points in the collection. Often the marginal mean of the field is not constant, so the mean is first modeled separately (such as with a linear model or a low-order polynomial) and then the detrended field can be fit with a stationary Gaussian process. Although the trend and the zero-mean spatial process are modeled separately, they can be fit simultaneously to allow a trade-off between complexity in the trend and a larger-magnitude spatial process.

Isotropy further simplifies the structure of a stationary field by requiring that the covariance between any two points depend only on the distance between them. Thus the covariance between one point and another point one unit to the northwest is the same as the covariance between the first point and another one unit to the northeast, and in fact with any point exactly one unit away from it. And by the assumption of stationarity, the same covariance applies to all pairs of points exactly one unit apart.

Thus, for a Gaussian process $x(\mathbf{s})$ observed at a set of locations $\mathbf{s}_1, \ldots, \mathbf{s}_n \in \mathcal{S}$, we can write its distribution in terms of its mean function $\mu(\mathbf{s})$ and its covariance function $C(\mathbf{s}_i, \mathbf{s}_j)$ as

$$\mathbf{x} \sim MVN\left(\boldsymbol{\mu}, \boldsymbol{\Sigma}\right),$$

where $\mathbf{x} = (x(\mathbf{s}_1), \ldots, x(\mathbf{s}_n))$, $\boldsymbol{\mu} = (\mu(\mathbf{s}_1), \ldots, \mu(\mathbf{s}_n))$, and $\boldsymbol{\Sigma}$ is the variance-covariance matrix with elements $C(\mathbf{s}_i, \mathbf{s}_j)$. With the typical assumptions of stationarity and isotropy, we can simplify this by letting $\mu(\mathbf{s}_i) = \mu$ for all i, and defining the covariance matrix such that

$$C(\mathbf{s}_i, \mathbf{s}_j) = \frac{1}{\lambda}\rho(d),$$

where λ is the precision (the inverse of the variance, which is easier to work with in a Bayesian context) and $\rho(d)$ is the correlation function that gives the correlation between any two points that are a distance of d units apart. Here we will simplify things by assuming a zero-mean process, so $\mu = 0$, but this is easily generalized to other mean functions, and lower-order polynomial functions of location are often used.

In most problems, $\mathcal{S} = \mathbb{R}^p$ and Euclidean distance is used as the distance metric. Other distance metrics are possible, and most are based on Euclidean distance on a transformation of $\mathcal{S}$. Such transformations are typically rotations and dilations (Isaaks and Srivastava, 1989, Chapter 16). More generally, spatial deformations can be used (Sampson and Guttorp, 1992; Schmidt and O'Hagan, 2003).

Since every covariance matrix must be nonnegative definite, the covariance function $C(\cdot, \cdot)$ must also be nonnegative definite. Typically one of several parametric forms is chosen for the correlation function, $\rho(d)$. Popular examples include the

spherical correlogram, $\rho(d) = \left(1 - \frac{3}{2}d + \frac{1}{2}d^3\right) I_{\{0 \le d \le 1\}}(d);$

exponential correlogram, $\rho(d) = e^{-d};$

Gaussian correlogram, $\rho(d) = e^{-d^2};$ and

Matérn class, $\rho(d) = \left[(d/2)^\nu 2K_\nu(d)\right]/\Gamma(\nu),$

where $I_{\{0 \le d \le 1\}}(d)$ is the indicator function, which is one when $0 \le d \le 1$ and zero otherwise, K_ν is a modified Bessel function of the second kind, and ν is a smoothness parameter (see Abramowitz and Stegun, 1964). The spherical correlogram is sometimes preferred because of its compact support. Stein (1999) presents a case for why the Matérn class (Matérn, 1986) should be preferred, particularly from a theoretical standpoint. However, from a more practical viewpoint, the smoothness parameter ν can be difficult to fit, particularly in the context of inverse problems (e.g., Chapter 16). The exponential and Gaussian correlograms are special cases of the Matérn class, with $\nu = 1/2$ and the limit as $\nu \to \infty$, respectively. We will typically use one of these simpler forms, such as the Gaussian correlogram, in this book.

Additional parameters can be used to specify the marginal variance and the range of spatial dependence so that $C(\mathbf{s}_i, \mathbf{s}_j) = \theta_1 \rho(d/\theta_2)$; e.g., the two-parameter Gaussian covariogram would be

$$C(d) = \theta_1 \exp\left[-\left(\frac{d}{\theta_2}\right)^2\right]. \qquad (2.4)$$

Figure 2.2 shows the correlogram (left column) and a random realization (right column) for the spherical, exponential, and Gaussian correlation functions.

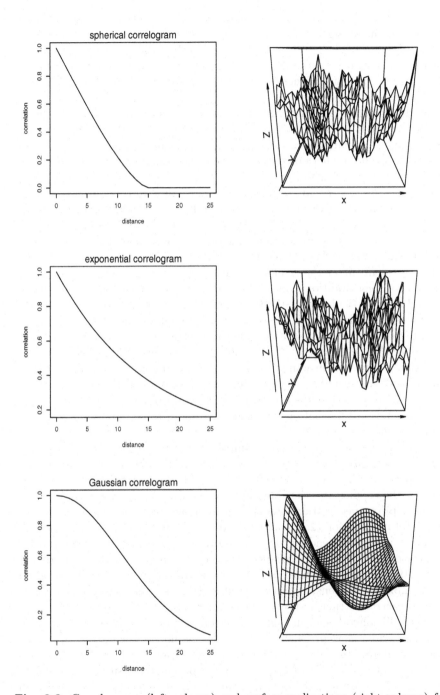

Fig. 2.2. Correlograms (left column) and surface realizations (right column) for spherical, exponential, and Gaussian correlation functions.

For this figure, θ_2 has been set to 15 for each plot. Note that the Gaussian function produces much smoother realizations than the other two functions. The smoothness of the realizations is related to the behavior of the correlation function near zero (Stein, 1999). Further details on choosing or modeling a covariance or correlation function can be found in Cressie (1993), Stein (1999), or Banerjee et al. (2003).

In the correlation structures above, the processes interpolate the data, passing through all data points and predicting continuously in between observations. In many cases, one wants to allow more flexibility in fitting, not requiring the predictions to match the data points exactly. Such a fit can then be a smoother function. Such a model can arise philosophically in one of two ways. First, one could consider a standard additive random noise model for the observations. Alternatively, one could consider additional variability as coming from a different small-scale process. This latter interpretation was the original motivation for the introduction of the nugget term (Matheron, 1963; Cressie, 1993). These two approaches can be shown to be mathematically equivalent under the right formulation. It is also worth noting that the inclusion of some additional variability can improve the numerical stability of the problem, particularly for the Gaussian correlation function (Neal, 1997).

In most spatial problems, the covariance parameters can be estimated from the data (although improper priors can cause problems in the Bayesian setting; see Berger, De Oliveira, and Sansó, 2001, for more discussion of noninformative priors). However, in some problems where no direct measurements of the spatial field are available, such as the permeability example in Chapter 16, most of these parameters will need to be chosen a priori, as the data do not contain sufficient information for the estimation of these parameters (see, for example, Oliver et al., 1997).

In the Bayesian setting, some simple cases can be dealt with in closed form. For example, if a covariance function is taken as known, the mean function is parameterized as a linear expansion, and a conjugate (normal) prior is used for the coefficients, then the posterior is available in closed form. With the right prior specification, the case of unknown marginal variance (the θ_1 parameter in Equation (2.4)) can be handled by using an inverse-gamma prior for the variance, also leading to a closed-form posterior and analytic forms for predictions and predictive variance. More details are available in Hjort and Omre (1994). We omit those details here because our focus tends to be on more complex situations, in particular either the case where additional aspects of the covariance structure need to be estimated (e.g., θ_2 in Equation (2.4)) or where the likelihood is significantly more complex (such as with the computer model for water flow in Chapter 16), and thus no fully conjugate specification is available. In such cases, posterior estimation needs to be done with Markov chain Monte Carlo methods. Some parameters (such as the coefficients for the mean and the marginal variance) can typically be treated with conditionally conjugate priors, allowing for Gibbs updates during MCMC. Other parameters (such as the range parameter θ_2) can only be updated via Metropolis-Hastings steps. When covariance parameters other than the marginal variance (θ_1) are unknown, the full covariance matrix must be inverted at each MCMC iteration, which can be computationally expensive as well as numerically unstable. The instability is a result of the matrix being close to singular, and the problem is exacerbated by smoother covariance functions, particularly the Gaussian covariogram. An important computational device is the Cholesky decomposition, which not only helps with numerical stability itself but can also be used with pivoting, allowing the rearrangement of the covariance matrix so that the focus of the fitting can be on the more important elements (see, for example, Lee et al., 2002).

This framework can be extended to settings with a spatially correlated multivariate response by employing methods such as cokriging (Ver Hoef and Barry, 1998; Wackernagel, 1998) or coregionalization (Schmidt and Gelfand, 2003; Gelfand et al., 2004).

An alternative specification for Gaussian processes uses a convolution approach, which is discussed in detail in Chapter 4. Using convolutions has certain computational benefits, and it also allows for a natural mechanism for modeling at different resolutions.

While not the focus of this book, it should be noted that a number of methods are available for modeling nonstationary processes, although these are typically limited to use on relatively smaller datasets. Examples include Haas (1990), Sampson and Guttorp (1992), Damian et al. (2001), and Schmidt and O'Hagan (2003). The convolution approach has also proved useful in modeling nonstationarity (Barry and Ver Hoef, 1996; Higdon et al., 1999; Fuentes and Smith, 2001), and Paciorek (2003) found an elegant extension that allows a number of expressions to be obtained in closed form. Kim et al. (2005) and Gramacy and Lee (2006) handle nonstationarity through partitioning approaches.

3

Illustrative Example

We present here a preliminary analysis of a dataset that is used in many parts of the book as an illustrative example for different multiscale modeling approaches. The dataset is the monthly flow of the Fraser River in Canada from January 1913 to December 1990. This dataset was previously analyzed by McLeod (1994) and is available from StatLib at

http://lib.stat.cmu.edu/datasets/fraser-river

Figure 3.1(a) presents the plot of the log of the mean monthly flows of the Fraser River from January 1913 to December 1990, as well as the series of its annual averages. The presence of seasonality in the monthly series is quite obvious. Also, it is evident from Figure 3.1(a) that there is strong temporal dependence between the annual averages.

During an exploratory analysis, we verified that the seasonality can be well explained by the first, fourth, and fifth harmonics. Figure 3.1(b) shows the plot of the monthly residuals after extracting the overall mean and the seasonality. Figure 3.2 shows the autocorrelation and partial autocorrelation functions of the monthly residuals. The autocorrelation function decays fairly slowly and thus suggests a long memory type process at the monthly level.

In contrast, Figure 3.3 shows the autocorrelation and partial autocorrelation functions of the annual series, strongly suggesting an AR(1) process for the annual level of aggregation. Thus, it seems that there are strong annual level dynamics that are reasonably well explained by an AR(1) process. Moreover, it seems that this annual level dynamics induces the long memory type of behavior at the monthly level. The annual level behavior is probably the result of large-timescale climate dynamics that impact water and snow precipitation and thus the flow of the river.

This example points to the necessity of having approaches for the analysis of processes with relevant dynamics at different levels of resolution. Chapter 11 presents the construction of explicit multiscale time series models from coarser to finer resolution levels. These models have the ability to describe processes that have dynamics at several time-resolution levels. In particular, Section 11.6.1 presents an analysis of the Fraser River with a multiscale time series model. Chapter 14 presents the computational construction of implicit multiscale models through the use of Metropolis coupling. This approach is illustrated on the Fraser River dataset in Section 14.2.1.

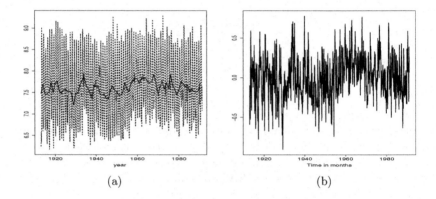

(a)   (b)

Fig. 3.1. Fraser River. (a) Log of the mean monthly flows (dotted) and annual averages (solid). (b) Monthly residuals after extracting the seasonality and the mean.

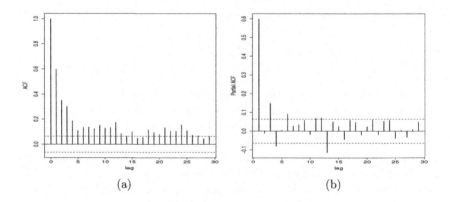

(a)   (b)

Fig. 3.2. Fraser River. (a) Autocorrelation and (b) partial autocorrelation functions of monthly residuals.

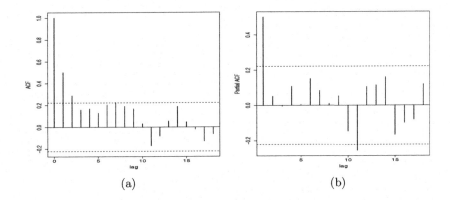

(a)   (b)

Fig. 3.3. Fraser River. (a) Autocorrelation and (b) partial autocorrelation functions of the annual series.

Part II

Convolutions and Wavelets

Convolution and Wavelet Overview

The two methods in this section, convolutions and wavelets, are grouped together because they are both multiscale decomposition methods. They are quite different in flavor from the explicit multiscale models of Part III (which make up the bulk of this book), as well as from the implicit multiscale models in Part IV. Hence they have received their own section. We include their part first, as some of the other methods in this book build upon their ideas. The two methods in this part are rather different from each other as well, both in mechanics and applicability. Yet they share an underlying basic conceptual approach.

The key idea is to build a model using multiple levels of components. This is essentially a hierarchical basis decomposition. Many nonparametric regression techniques rely on using an infinite basis set that spans the space of continuous functions. For example, the set of all polynomials spans the space of continuous functions, as do particular sets of orthogonal polynomials (e.g., Legendre, Laguerre, Hermite). Similarly, Fourier series regression relies on using a theoretically infinite set of sine and cosine functions to span the space of interest. For any of these basis function approaches, in practice only a finite number of basis terms are used, and using enough terms gives a close enough approximation. Additional discussion on families of nonparametric regression methods can be found in Lee (2004, Chapter 2). Similarly, both convolutions and wavelets can be thought of as using a basis expansion. However, the basis functions are instead ordered hierarchically (as opposed to by the order of the polynomial, for example), with larger-scale components being considered first and successively smaller-scale components added sequentially. Models can thus be built using multiple levels of resolution. At some point, the hierarchy is truncated when it is deemed that a sufficient number of kernels or wavelets have been included.

This type of approach gives a way to model different aspects of a dataset, with larger-scale components modeling overall trends and smaller-scale components modeling finer details. In some cases, this additional interpretability of the models may be beneficial. Done correctly, it may be possible to match the modeling scales to physically meaningful scales, particularly with the convolution approach, where the scales can be chosen more arbitrarily. In other cases, the scales should be seen more as mathematical constructions chosen for computational convenience, and attempting to ascribe physical interpretability may not make sense. As always, such ideas will be highly dependent on the particular application.

We do want to note that, for both convolutions and wavelets, most uses of these methods in the literature are not explicitly multiscale. Most frequently, these methods are used as nonparametric regression techniques, and they just happen to organize their basis components hierarchically. Both approaches have seen great successes in a large variety of applications. Wavelets in particular have found widespread popularity. However, we will skip over large sections of the literature in the following chapters so that we can focus on the aspects that are most relevant for explicit multiscale uses.

4

Convolution Methods

Standard Gaussian process models were introduced in Section 2.2. Such models gained popularity originally in geostatistical applications but have become widely used throughout spatial statistics. In this chapter, we will see how one can represent a Gaussian process via convolutions, which can have great computational advantages as well as providing a rather natural mechanism for moving to a multiscale model. Thus the latter section of this chapter will fully explore how to use the convolution representation to build a multiscale model.

4.1 Convolutions

A convenient way of obtaining a Gaussian process is given by convolving white noise with a smoothing kernel (Thiébaux and Pedder, 1987; Barry and Ver Hoef, 1996). Following Higdon (2002), let $w(\mathbf{s})$ be a white noise process (or Wiener process "derivative"; see, e.g., Priesley, 1981), $\mathbf{s} \in \mathcal{S}$, and let $k(\cdot; \phi)$ be a kernel, possibly depending on a low-dimensional parameter ϕ. Then a Gaussian process is obtained by

$$x(\mathbf{s}) = \int_{\mathbf{S}} k(\mathbf{u} - \mathbf{s}; \phi) w(\mathbf{u}) du = \int_{\mathbf{S}} k(\mathbf{u} - \mathbf{s}; \phi) dW(\mathbf{u}), \qquad (4.1)$$

where W is a Wiener process. The resulting process $x(\mathbf{s})$ has a covariance function that depends only on the displacement vector $\mathbf{d}_{s,s'} = \mathbf{s} - \mathbf{s}'$ for $\mathbf{s}, \mathbf{s}' \in \mathbf{S}$; i.e.,

$$
\begin{aligned}
C(\mathbf{d}_{s,s'}) = \operatorname{cov}(x(\mathbf{s}), x(\mathbf{s}')) &= \int_{\mathbf{S}} k(\mathbf{u} - \mathbf{s}; \phi) k(\mathbf{u} - \mathbf{s}'; \phi) du \\
&= \int_{\mathbf{S}} k(\mathbf{u} - \mathbf{d}_{s,s'}; \phi) k(\mathbf{u}; \phi) du.
\end{aligned}
$$

As noted in Kern (2000), $C(\mathbf{d}_{s,s'})$ is the convolution of the kernel with itself. If $\mathbf{S}$ is $\mathbb{R}^p$ and $k(\mathbf{s})$ is isotropic, then $x(\mathbf{s})$ is also isotropic, with covariance function $C(\mathbf{d}_{s,s'})$, which depends only on the magnitude d of $\mathbf{d}_{s,s'}$. In this case, there is a one-to-one relationship between the smoothing kernel $k(d)$ and the covariogram $C(d)$, provided either $\int_{\mathbb{R}^p} k(\mathbf{s})d\mathbf{s} < \infty$ and $\int_{\mathbb{R}^p} k^2(\mathbf{s})d\mathbf{s} < \infty$ or $C(\mathbf{s})$ is integrable and positive definite (Thiébaux and Pedder, 1987). In particular, for a given $C(\cdot)$, $k(\cdot; \phi)$ can be obtained as the inverse Fourier transform of the square root of the spectral density of C. This relationship is based on the convolution theorem for Fourier transforms; more details can be found in Barry and Ver Hoef (1996). An example of a flexible family of isotropic kernels is obtained from the Matèrn class of isotropic correlations (see, for example, Stein, 1999), which is known to correspond to processes with widely different degrees of smoothness. The spectral density of the correlation function in the $\mathbb{R}^2$ Matèrn class, with range $\lambda > 0$ and smoothness $\nu > 0$, is given by $f(\Upsilon) \propto 1/(\lambda^2 + \Upsilon^2)^{\nu/2+1}$. The corresponding kernel is the inverse Fourier transform of $1/(\lambda^2 + \Upsilon^2)^{\nu/4+1/2}$, which is proportional to

$$(\lambda s)^{\nu} \mathcal{K}_{\nu}(\lambda s), \quad \lambda > 0, \nu > 1, \tag{4.2}$$

where $\mathcal{K}_{\nu}$ is the modified Bessel function of the second kind of order ν (Abramowitz and Stegun, 1964).

In the rest of this chapter, we focus the discussion and examples on processes in one dimension. However, the results are readily applied to higher dimensions as well. In practice, the need to model the background process tends to limit the application of the convolution approach to three or fewer dimensions as, with more dimensions than that, the curse of dimensionality starts to kick in and it is typically more efficient to model the Gaussian process directly.

A discrete approximation of Equation (4.1) can be obtained by fixing a finite number of evenly spaced points, say $s_1, \ldots, s_M$, and giving

$$x(s) \approx \sum_{i=1}^{M} k(s_i - s; \phi)w(s_i), \tag{4.3}$$

where $w(\cdot)$ is white noise (Higdon, 2002). (Note that the kernel may need to be rescaled to be comparable with Equation (4.1)). Figure 4.1 shows this approximation in action. Each of the three plots shows a process produced by smoothing a background process of nine points (at $\mathbf{s} = (1, \ldots, 9)$) drawn independently from Gaussian distributions, $w(s_i) \overset{iid}{\sim} N(0, 1)$, and smoothed with a standard Gaussian kernel. The solid dark line shows the approximate Gaussian process, and the light curves show the individual kernels, with the smaller vertical black lines denoting the background process. The three plots differ only in that the background process has been regenerated each time. Note that a smooth curve results from the convolution of a discrete process and that the smoothness is a function of the choice of kernel, so all three curves have similar smoothness and covariance structure because they were generated using the same kernel. (The initial downward trend in all three is a coincidence of the particular set of random numbers drawn.)

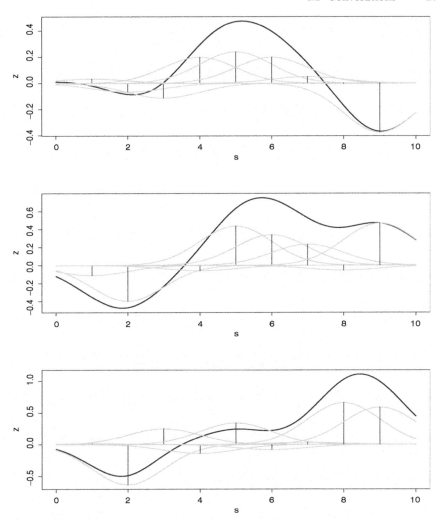

Fig. 4.1. Examples of Gaussian processes built from discrete convolutions. The grey lines show the individual kernels, and the dark lines are the resulting processes.

As mentioned earlier, the choice of kernel is the primary determinant of the properties of the resulting process. Figure 4.2 demonstrates the effect of changing the kernel width on the resulting curve. The background process is the same as in the middle plot of Figure 4.1, but the standard deviation of the Gaussian kernels has been changed. (Note that even though the background process is the same, the kernel heights shown in the figure vary because the kernels are normalized to have unit area.) Whereas it was 1.0 in Figure 4.1, in Figure 4.2 it is (from top to bottom) 0.4, 0.6, and 2.0. Notice that smaller kernels produce more wiggly curves, while larger kernels give rise to extremely

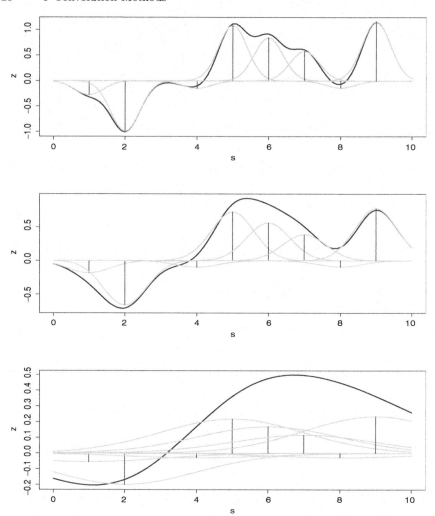

Fig. 4.2. Effect of kernel width on the resulting Gaussian process approximation. The kernel shape and discrete process values are the same as in the middle graph of Figure 4.1, but the kernel width is smaller in the top two graphs and larger in the bottom one.

smooth curves. Just as with standard approaches to Gaussian processes (see the last paragraph of Section 2.2), if the data allow, the width of the kernel can be treated as a parameter and fit from the data along with the values of the background process. In other cases, the kernel parameters may need to be chosen a priori.

The choice of kernel is not restricted to the Gaussian kernels used in Figures 4.1 and 4.2. Although they are frequently a convenient choice for

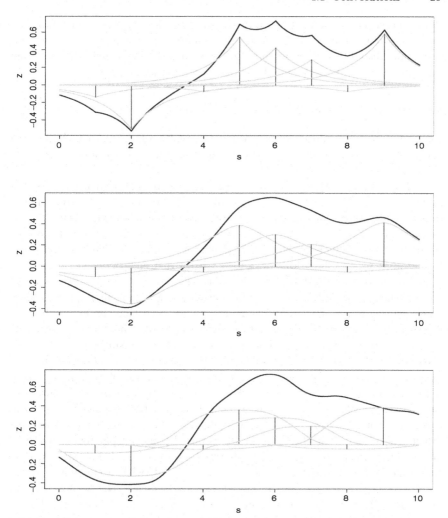

Fig. 4.3. Effect of kernel shape on the resulting Gaussian process approximation. The discrete process values are the same as in the middle graph of Figure 4.1. The kernels are for an exponential variogram, for a Matérn process with shape parameter 1.5, and the tricube kernel.

relatively smooth functions, they can be too smooth for other applications. Examples of other choices of kernels on the same background process are shown in Figure 4.3. The top plot shows the kernels that would lead to the exponential variogram, suitably standardized. Note that, in a real application, more background points would be needed for nonsmooth kernels such as these in order to produce a good realization; here we keep the number of background points fixed for comparison with the other plots. The middle plot shows the kernels that result in a process in the Matérn class with shape parameter 1.5, a curve of intermediate smoothness between the exponential and Gaussian variograms. The bottom plot uses the tricube kernel, a local kernel that produces a different sort of smoothness. This kernel is given by

$$k(s) = \left(1 - \frac{|s|^3}{\lambda^3}\right)^3 I[|s| < \lambda],$$

where λ is a range parameter affecting the width of the kernel. Additional discussion of the relationship between the kernel and the process is given in Kern (2000).

A lesser issue is the choice of spacing for the background process. If the points are too far apart, then the kernels will not be able to smooth the process effectively. But once a reasonable saturation has been reached, there are strongly diminishing returns from adding additional background locations. A rule of thumb given by Higdon is that a good compromise is to have the background process locations on a grid with a distance between the nearest points equal to the standard deviation of the kernels (although more localized kernels such as that leading to exponential variograms will require a more dense grid). Figures 4.4 and 4.5 show the results of fitting the log of the first three years of the Fraser River data (from Chapter 3) with Gaussian kernels. Here the values of the background process are fit using Markov chain Monte Carlo methods (see the next paragraph for a discussion of model fitting). In each case, Gaussian kernels with a standard deviation of four months are used, while the spacing of the background process changes. In Figure 4.4, the plots from top to bottom use background grid points placed every two, three, and four months, respectively. Note that there is almost no difference in the fitted curves for the three plots. Figure 4.5 shows spacings of five, six, and seven months, respectively, from top to bottom, and one can see that the curves start to differ from those in Figure 4.4, with a spacing of seven months basically destroying the model. In this last case, the kernels are too far apart. While in many cases using too few kernels results in an overly lumpy fit (overfitting certain regions and ignoring others), here the periodic nature of the data results in the best kernel fit being close to constant. In summary, spacing the background points equal to the standard deviation of the kernel is usually sufficient, in that little additional gain is to be had by using more background points. One can sometimes get away with using even fewer, but using too few leads to degradation in the fit. Also note that it is helpful for the background grid to be slightly larger than the space of the observed data, so that there is a buffer of at least one point in each direction outside of the convex hull of observed points.

Fitting a convolution-based process to data is a matter of finding the values of the background process points, and often also fitting some or all of the other parameters in the model (parameters defining the kernel, the variance of the error terms, the variance of the background process, and any parameters in the mean function of the process). Higdon (2002) provides a

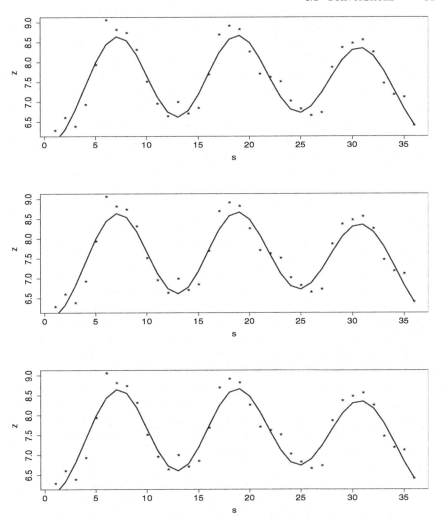

Fig. 4.4. Effect of kernel spacing on the resulting Gaussian process approximation. The background points are spaced 2, 3, and 4 units apart in the plots.

worked-out example with code in R or S (which can also be found in the technical report version, Duke Institute of Statistics and Decision Sciences working paper 01-03). We also provide basic code at the Web site for this book (http://www.ams.ucsc.edu/~herbie/multiscale). If all of the other parameters are taken as known, then the problem is equivalent to fitting a mixed-effects model and can be done via restricted maximum likelihood estimation (REML), such as with the E-M algorithm (Dempster et al., 1984). In a fully Bayesian setting, the posterior is not generally available in closed form, and Markov chain Monte Carlo methods are necessary. MCMC theoretically

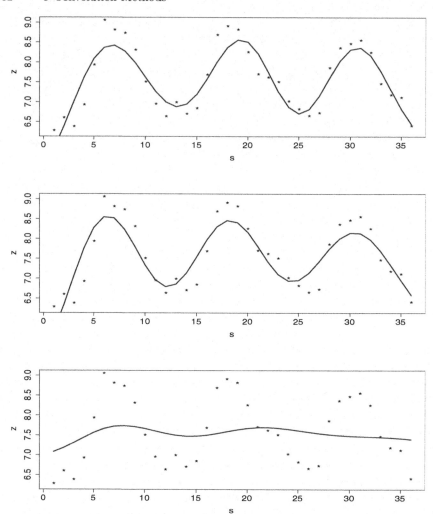

Fig. 4.5. Effect of kernel spacing on the resulting Gaussian process approximation. The background points are spaced 5, 6, and 7 units apart in the plots.

also allows fitting any number of other parameters in the model. In practice, the data may be limited in the amount of information that they provide, and some of the parameters may need to be fixed a priori. Note also that some care must be taken with the priors for the kernel parameters, in that improper priors may lead to improper posteriors (Berger et al., 2001). Using proper priors for the kernel parameters avoids this problem.

This finite representation has many computational advantages, as only a relatively small number of sites s_i are necessary for a close approximation, saving much computing time. This reduction in dimension is also useful for a variety of theoretical reasons; for example, many inverse problems are drastically ill-conditioned, and the dimension reduction of the parameter space greatly aids in the ability to conduct inference. Note that this approach still produces a continuous process, even though the underlying process is discrete.

The class of resulting spatial processes can be enlarged by either allowing the kernel to vary spatially (Higdon et al., 1999) or by changing the background process. Examples of the latter include background processes of Gaussian processes (Fuentes and Smith, 2001), Markov random fields (Lee et al., 2005), and non-Gaussian processes (Ickstadt and Wolpert, 1999). Some similar ideas have also appeared in the spatio-temporal literature (Wikle et al., 1998; Calder et al., 2002; Stroud et al., 2001).

4.2 Multiscale Convolutions

The convolution approach to generating Gaussian processes is easily adapted to model a process at multiple resolutions (Higdon, 2002). The process is taken to be a sum of separate component processes at different resolutions, accumulating from coarse to fine. Finer levels use kernels of smaller width on a larger number of background grid points. To be explicit, the multiresolution process with L levels is given by

$$x(s) = \sum_{l=0}^{L-1} x_l(s) = \sum_{l=0}^{L-1} \sum_{i=1}^{m_l} k_l(s_i - s; \phi) w_l(s_i), \qquad (4.4)$$

where $x_l(s)$ is the lth resolution process, and each of these processes is generated from a convolution of kernels k_l on an appropriate resolution background process $\mathbf{w}_l$. As before, a mean term or mean function can be added in as well.

If the process is observed on multiple scales, then the terms accumulate only to the scale being modeled. For example, if observations occur at three scales, then the coarsest is modeled as just $x_0(s)$, while the intermediate process is modeled with $x_0(s) + x_1(s)$ and the finest with $x_0(s) + x_1(s) + x_2(s)$, i.e.,

$$y_l(t_j) = \mu_l(t_j) + \sum_{q=0}^{l} \sum_{i=1}^{m_q} k_q(s_i - t_j; \phi) w_q(s_i) + \varepsilon_l(t_j), \qquad (4.5)$$

where $y_l(t_j)$ is an observation of the process at the lth scale, $j \in \{1, \ldots, n_l\}$, $t_j \in \mathcal{S}_l$ (the same space as the s_i, which often does not depend on scale), $\mu_l(s)$ is the mean process at scale l (taken here as a constant at each scale, but this could be easily generalized), and $\varepsilon_l(t_j)$ is iid mean zero Gaussian error whose variance could depend on the scale. Additional scales can be added for computational or modeling reasons at any point, and they accumulate appropriately.

If the process is only observed at a single scale, this formulation can still be useful for explicitly modeling large-scale trends separately from small-scale activity, as with the examples below.

A typical structure would be for $\mathbf{x}_{l+1}$ (the next finer resolution than $\mathbf{x}_l$) to have kernels k_{l+1} with effective width (e.g., standard deviation for a normal) one half that of k_l and to use twice as many grid cells in each dimension for $\mathbf{w}_{l+1}$ as compared to $\mathbf{w}_l$. Clearly this is most practical for only a small number of scales in a relatively small number of dimensions, as the computational requirements can quickly grow out of hand. Computational restrictions or information about the problem of interest may suggest other designs.

Choosing appropriate priors becomes more difficult than in the single-resolution setting, where one may have a good idea about what the process should look like. It may be harder to have good intuition about a multiresolution process (although sometimes the problem will have useful information to incorporate). Thus it can be helpful to have default priors that work reasonably well in a variety of situations. Following Higdon (2002), we have found that vaguely informative priors on a small hierarchy work well in practice. Both the background processes and the observational error are taken to be Gaussian with unknown precision (the reciprocal of the variance, which helps simplify the structure). The gamma distribution is conditionally conjugate for the precision of a normal, and a reasonable vague specification for a gamma is with shape parameter 1 and mean 200 (which is an exponential with rate 0.005). With more information, other gamma parameters or other distributions can be used. These parameters can also vary by resolution. In the absence of such information, these values provide a good starting point. Thus the full hierarchical model is

$$
\begin{aligned}
\varepsilon_l(t_j) &\sim N(0, \lambda_y^{-1}) & l &= 0, \ldots, L-1; \ j = 1, \ldots, n_l \\
w_l(s_i) &\sim N(0, \lambda_w^{-1}) & l &= 0, \ldots, L-1; \ i = 1, \ldots, m_l \\
\lambda_w &\sim \Gamma(1, 0.005) \\
\lambda_y &\sim \Gamma(1, 0.005).
\end{aligned}
$$

This setup is easily expanded to allow different hyperparameters on the λ parameters at different scales if desired. If a mean parameter is included, it can either be given a flat prior or the same prior as the background process (which helps to simplify the computations).

As with the single-resolution convolution model, the background process can be fit via either REML or MCMC, and the full posterior including unknown parameters can be found with MCMC methods (Higdon, 2002). We provide basic code for multiresolution modeling with convolutions at our Web site (http://www.ams.ucsc.edu/~herbie/multiscale). This code can serve as a template and is easily modified for a variety of more complex situations.

4.2.1 Synthetic Example

The example in this section is inspired by one from Higdon (2002) and is of a known process that is composed of two subprocesses that are harmonic with

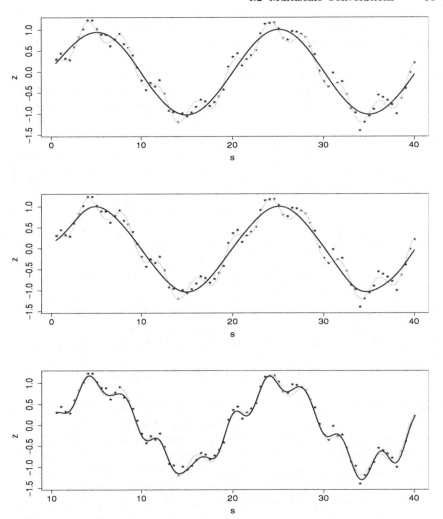

Fig. 4.6. Standard convolutions fit to synthetic data. The Gaussian kernels have standard deviation 5, 2, and 1. Dark lines show the fitted functions, light lines the true function.

different periods. Thus this example is of a process observed on a single scale but that can be modeled in a multiscale manner. In particular, the data are generated from the true function plus Gaussian noise

$$y(t_i) = \sin\left(\frac{2\pi t_i}{20}\right) + \cos\left(\frac{2\pi t_i}{4}\right) + \varepsilon(t_i)$$

for $t_i \in (1, 40)$ and $\varepsilon(t_i) \sim N(0, (0.1)^2)$. The truth is shown in Figures 4.6 and 4.7 as the solid grey line, with the generated data shown as the points scattered about the line. This truth is a combination of a lower-frequency smooth sine curve and a higher-frequency cosine curve, and the convolution approach provides a mechanism for capturing both of these pieces simultaneously. For comparison, first standard single-scale convolutions are fit to the data, with Figure 4.6 showing the results of fitting convolutions using Gaussian kernels with standard deviations 5, 2, and 1 (from top to bottom). The fitted curves are shown with dark solid lines. Note that there is little difference between the first two, even though only ten background points are used in the first while 22 are used in the second. The last plot uses 43 background points (on the unit locations from 0 to 42) and achieves a much finer resolution of fit, as the smaller kernels allow a closer fit. The first two choices of kernels are too smooth to fully model this particular dataset with only a single scale.

Figure 4.7 demonstrates the results of using multiscale convolution models. First, for comparison, the top plot again shows the fit with a single convolution with standard deviation one Gaussian kernels and also plots all of the individual kernels. The second plot shows a multiresolution model with Gaussian kernels of standard deviations 5 and 1 and also shows the individual kernels, with dashed lines for the larger kernels and dotted lines for the smaller ones. The third plot shows a multiresolution model with Gaussian kernels of standard deviations 3 and 1, and leaves out the individual kernels to reduce the clutter in the plot. First, note that all three fitted curves are very similar, so, as before, the larger-scale process can be modeled by a range of possible larger kernels. Now compare the individual kernels in the top two plots. Notice that in the second plot the larger kernels capture the larger motion of the sine curve, while the smaller kernels capture the finer motion of the cosine curve. The smaller kernels can be viewed as modeling the local deviations from the larger process. In contrast, in the top plot, the smaller kernels have to do double duty, modeling both the finer and coarser processes at the same time. If one really only cares about the finest scale, such as with interpolation or prediction, then the single-scale model will work just fine. But if one is trying to understand the underlying process at multiple scales (e.g., a process such as pollution that has both weekly and daily cycles), then the multiresolution convolution is more appropriate, as it can produce separate fits for the two separate cycles.

4.2.2 Fraser River Example

We now return to the ongoing Fraser River example (introduced in Chapter 3). While Figure 4.4 shows the results of a convolution process using only a single set of kernels, Figure 4.8 shows a multiresolution example. The top plot shows the fit from a multiresolution convolution using Gaussian kernels of standard deviations 4 and 1, with the kernels spaced on a grid of matching size (i.e., one background point every four units for the larger kernels, one every unit for the smaller kernels), with the inclusion of one additional background point on either end (to help eliminate edge effects). Notice that

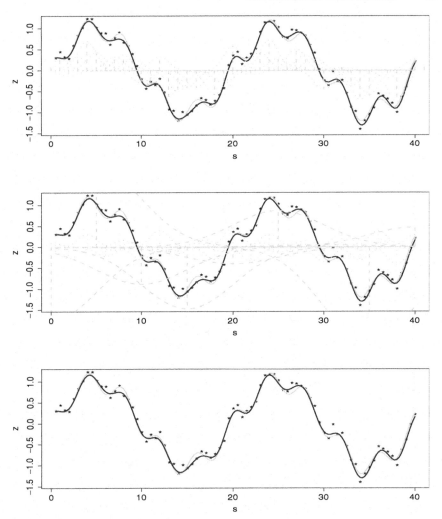

Fig. 4.7. Standard and multiresolution convolutions fit to synthetic data. The top plot shows a standard convolution with its Gaussian kernels of standard deviation 1. The middle plot shows a multiresolution fit with its kernels of sizes 5 and 1. The bottom plot shows just the fit for kernels of sizes 3 and 1. Dark lines show the fitted functions, light solid lines the true function.

the fit is slightly better than in Figure 4.4. For comparison, the bottom plot of Figure 4.8 shows a single-resolution fit using smaller kernels of standard deviation 2. Note that the fitted curves in the top and bottom plots are quite similar. The middle plot shows the decomposition of the multiresolution fit, with the fits due to the larger and smaller kernels graphed separately (with the smaller kernel effects being shown as deviations from the mean level). The

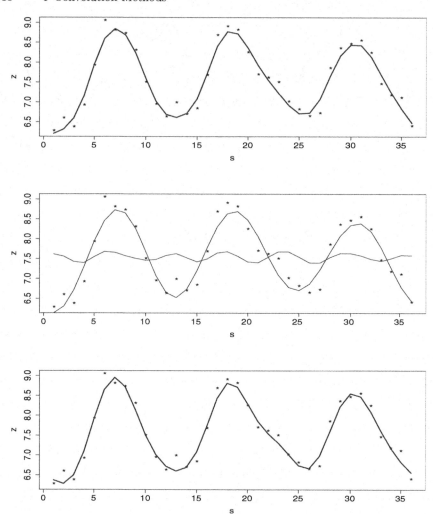

Fig. 4.8. Multiresolution and standard convolutions fit to the Fraser River data. The top plot shows a multiresolution fit with Gaussian kernels of sizes 4 and 1. The middle plot shows the separate fit components for the larger and smaller kernels. The bottom plot shows a standard convolution with Gaussian kernels of standard deviation 2.

annual trend is neatly captured by the larger (standard deviation 4) kernels, while a shorter-term, presumably seasonal, trend is captured by the smaller kernels. Thus, when interpretability at multiple scales is required, multiresolution convolutions provide a convenient representation.

5

Wavelet Methods

Wavelets are one of the most well-known multiscale techniques and have been extensively explored in statistical applications in the 1990s and the 2000s. Wavelets are powerful mathematical tools that have been used with success in the compression of images and in the analysis of signals. They decompose signals and images in components spatially localized at different levels of resolution known as wavelet coefficients.

Bayesian wavelet analysis is useful when the signal is subject to random error. In that case, the wavelet transform of the observations will provide noisy empirical wavelet coefficients. In Bayesian analysis, the use of a joint prior distribution for the wavelet coefficients and an appropriate loss function will lead to a Bayesian estimate that will shrink or threshold the empirical wavelet coefficients toward zero. The inverse wavelet transform of the estimated wavelet coefficients will provide a noise-free estimate of the signal.

In this book, we focus not only on multiscale decomposition methods such as wavelets but also on other multiscale statistical models. These other models are more suitable for time series prediction, for problems where data are available at different resolution levels, or when dealing with a natural multiscale stochastic process.

While Fourier analysis decomposes a signal into its frequency components, a wavelet analysis decomposes a signal into its time-scale (or time-frequency) components. Wavelet functions are obtained by translations and scalings of the so-called mother wavelet. The mother wavelet is a function that has some properties related to the very name *wavelet*. The first property, known as the admissibility condition, implies that the mother wavelet integrates to zero, and thus it is a wave. The second property, known as the regularity condition, implies that the mother wavelet has vanishing moments, and, as a consequence, it is well localized. A function with these properties is then a small wave – a wavelet. This chapter is intended to give an overview on wavelets. For more detailed information on wavelets, the reader is referred to the excellent books by Daubechies (1992), Mallat (1999), and Vidakovic (1999). A comprehensive overview of Bayesian wavelet methods is given in Müller and Vidakovic (1999a).

This chapter is organized as follows. Section 5.1 provides some notation and background information. Section 5.2 presents the continuous wavelet transformation. Section 5.3 introduces the scaling function. Section 5.4 presents discrete wavelets and the discrete wavelet transform. Bayesian nonparametric regression using wavelets is discussed in Section 5.5.

5.1 Background

Wavelets can be used to analyze functions $f(t)$ that belong to the set of finite energy functions $L^2(\mathbb{R})$; that is, $\int_{-\infty}^{+\infty} |f(t)|^2 dt < +\infty$.

In the definition of the wavelet transform, we will use inner products between functions, defined as $\langle f, g \rangle = \int_{-\infty}^{+\infty} f(t)\overline{g(t)}dt$, where $\overline{g(t)}$ is the complex conjugate of $g(t)$. Moreover, we will consider here the L^2 norm: $||f|| = \langle f, f \rangle$.

A wavelet is a function $\psi \in L^2(\mathbb{R})$ such that

- $\int_{-\infty}^{+\infty} \psi(t)dt = 0$;
- $||\psi|| = 1$.

The function ψ may be used to generate by translation and scaling operations a whole family of wavelets. For this reason, ψ is called the mother wavelet. More precisely, translating ψ by u and scaling it by s leads to

$$\psi_{su}(t) = \frac{1}{\sqrt{s}}\psi\left(\frac{t-u}{s}\right), \qquad (5.1)$$

where $u \in \mathbb{R}$ and $s \in \mathbb{R}^+$. Depending on whether $s < 1$ or $s > 1$, then ψ_{s0} will be a contracted or an expanded version of ψ, respectively.

The simplest and maybe most well-known wavelet is the Haar wavelet (Haar, 1910). The mother Haar wavelet is

$$\psi(t) = \begin{cases} 1, & 0 \leq t < 0.5, \\ -1, & 0.5 \leq t < 1, \\ 0, & \text{otherwise.} \end{cases}$$

Due to its blockiness, the Haar wavelet is not appropriate to approximate smooth functions. Fortunately, there are many other families of wavelets. In particular, the Daubechies wavelet family (Daubechies, 1992) has been extensively used in statistical applications because its members have compact support and can be chosen according to the smoothness of the function of interest.

5.2 Continuous Wavelet Transform

A continuous wavelet transform (CWT) can be used to decompose any finite energy function in its time-scale components, and an inverse continuous wavelet transform can be used to perfectly reconstruct that function. This ability mirrors that of the Fourier transform for the analysis of stationary signals.

The CWT of f based on ψ computed for translation u and scale s is defined as

$$W_f^{\psi}(s,u) = \langle f, \psi_{su} \rangle = \int_{-\infty}^{+\infty} f(t)\overline{\psi_{su}(t)}dt. \tag{5.2}$$

When applied to a one-dimensional function, the CWT is a two-dimensional function and thus is a highly redundant transform. We will go back to this redundancy later.

Let us consider the admissibility condition (Calderón, 1964; Grossmann and Morlet, 1984)

$$C_{\psi} = \int_{-\infty}^{\infty} \frac{|\hat{\psi}(w)|^2}{|w|}dw < \infty,$$

where $\hat{\psi}(w) = \int_{-\infty}^{\infty} \psi(t)e^{-iwt}dt$, the Fourier transform of ψ, measures the amount of oscillation of ψ at the frequency w. The admissibility condition implies that $\hat{\psi}(0) = 0$. But $\hat{\psi}(0) = \int_{-\infty}^{\infty} \psi(t)dt$ and thus the mother wavelet integrates to zero.

If the mother wavelet $\psi(t)$ satisfies the admissibility condition, then any function $f \in L^2(\mathbb{R})$ satisfies (Calderón, 1964; Grossmann and Morlet, 1984)

$$f(t) = \frac{1}{C_{\psi}} \int_0^{+\infty} \int_{-\infty}^{\infty} \frac{1}{s^2} W_f^{\psi}(s,u)\psi_{su}(t)duds;$$

that is, the CWT based on $\psi(t)$ is invertible. Moreover, under the admissibility condition, a continuous wavelet transform conserves energy:

$$\int_{-\infty}^{\infty} |f(t)|^2 dt = \frac{1}{C_{\psi}} \int_0^{+\infty} \int_{-\infty}^{\infty} \frac{1}{s^2} |W_f^{\psi}(s,u)|^2 duds.$$

As mentioned before, the continuous wavelet transform is highly redundant. That is because two different wavelets ψ_{su} and $\psi_{s_0 u_0}$ are in general nonorthogonal. The correlation between ψ_{su} and $\psi_{s_0 u_0}$ is measured by the so-called reproducing kernel $K(s, u, s_0, u_0) = \langle \psi_{su}, \psi_{s_0 u_0} \rangle$. It is easy to show that

$$W_f^{\psi}(s_0, u_0) = \frac{1}{C_{\psi}} \int_0^{+\infty} \int_{-\infty}^{\infty} \frac{1}{s^2} K(s, u, s_0, u_0) W_f^{\psi}(s,u)duds.$$

The preceding equation is known as the reproducing kernel equation.

As the CWT is redundant, one can reduce the computational effort by using discrete values of u and s and still have an invertible transformation. The most used discretization is defined by

$$u = 2^{-l}k, s = 2^{-l}, k, l \in \mathbb{Z}.$$

As discretizations coarser than this discretization do not lead to an invertible transformation, the discretization above is known as critical sampling. When using critical sampling, the wavelet at scale l and translation k can be rewritten as

$$\psi_{lk}(t) = 2^{l/2}\psi\left(2^l t - k\right). \tag{5.3}$$

For a positive integer l, ψ_{l0} is a contracted version of the mother wavelet ψ, whereas for negative integer l, ψ_{l0} is an expanded version of the mother wavelet. Thus, larger l's will correspond to ψ_{lk} being able to capture localized high-frequency details, whereas smaller l's will correspond to ψ_{lk} being able to capture global low-frequency behavior.

Under mild conditions, $\{\psi_{lk}(t), l, k \in \mathbb{Z}\}$ will be an orthogonal basis for $L^2(\mathbb{R})$ (see Vidakovic, 1999). In that case, any function $f \in L^2(\mathbb{R})$ can be expanded as

$$f(t) = \sum_{l,k \in \mathbb{Z}} \psi_{lk}(t) W_f^{\psi}(2^{-l}, 2^{-l}k). \tag{5.4}$$

Henceforth, we consider orthogonal wavelets with critical sampling.

5.3 Scaling Function

Very often in practice the wavelet transform of a function f will be computed only for scales l larger than a given threshold l_0. In order to recover the function f, it will be necessary to have a coarser version of f at scale l_0. This coarser version will be obtained with the help of the so-called scaling function.

The scaling function ϕ, also known as the father wavelet, can be seen as an aggregation of wavelets at scales l coarser than 0. Let us define

$$\phi_{lk}(t) = 2^{l/2}\phi(2^l t - k).$$

The scaling function ϕ is such that $\{\phi_{l_0 k}, \psi_{lk}, l > l_0, k \in \mathbb{Z}\}$ is an orthogonal basis for $L^2(\mathbb{R})$.

Let us define

$$L_f^{\phi}(l, k) = \langle f, \phi_{lk}\rangle.$$

The coarser approximation of f at scale l_0 is

$$\sum_{k \in \mathbb{Z}} \phi_{l_0 k}(t) L_f^{\phi}(l_0, k).$$

Thus, the wavelet expansion (5.4) becomes

$$f(t) = \sum_{k \in \mathbb{Z}} \phi_{l_0 k}(t) L_f^{\phi}(l_0, k) + \sum_{l \geq l_0} \sum_{k \in \mathbb{Z}} \psi_{lk}(t) W_f^{\psi}(2^{-l}, 2^{-l} k). \qquad (5.5)$$

For the Haar wavelet, the corresponding scaling function is:

$$\phi(t) = \begin{cases} 1, & 0 \leq t < 1, \\ 0, & \text{otherwise.} \end{cases}$$

When using the Haar wavelet, the coarser version of a function f at resolution level l_0 will be a step function with step lengths equal to 2^{-l_0} and with step value equal to the mean of the function in the corresponding interval. We can then consider a succession of coarser versions of f at different resolution levels, as illustrated in Figure 5.1 for the function HeaviSine $f(t) = 4 \sin(4\pi t) - \text{sign}(t - 0.3) - \text{sign}(0.72 - t)$ considered by Donoho and Johnstone (1994). Figure 5.1 shows the original function and coarser versions at resolution levels 3 through 7. At resolution level 3, the approximation is very rough, but as the resolution level increases more details are added, and at resolution level 7 the approximation is very close to the original function. The approximations at different resolution levels may highlight different aspects of the function of interest. For example, the approximation at resolution level 3 highlights the large-scale features of the HeaviSine function, such as the existence of two major valleys and two major hills. Complementarily, the approximation at resolution level 7 highlights small-scale features of the function, such as the discontinuities and one peak that was hidden in the level 3 analysis.

5.4 Discrete Wavelets and the Discrete Wavelet Transform

In practice, the function f will not be observed continuously but instead will be sampled within a given interval of the real line with measurement error. In this section, we consider the case without measurement error, and in Section 5.5 we consider the case with measurement error. As often found in practice, we assume here that the function is sampled at equally spaced intervals and there are $n = 2^L$ sampled points. The discrete wavelet transform will consider wavelet and scaling functions sampled at those same points. As in the wavelet analysis for continuous functions, the discretized function will be expanded in terms of the discretized wavelet and scaling functions.

Without loss of generality, assume that the observations are taken on the interval $[0, 1]$. Denote by $t_i = (i - 1) 2^{-L}$ and by $f_i = f(t_i)$, $i = 1, \ldots, 2^L$ the sampling points and the sampled values of the function f.

As the sampled interval is finite, there is a lower limit for the coarsest possible scale, which we assume to be $l = 0$. Moreover, as the number of sampled points is 2^L, the finest possible scale is $l = L - 1$.

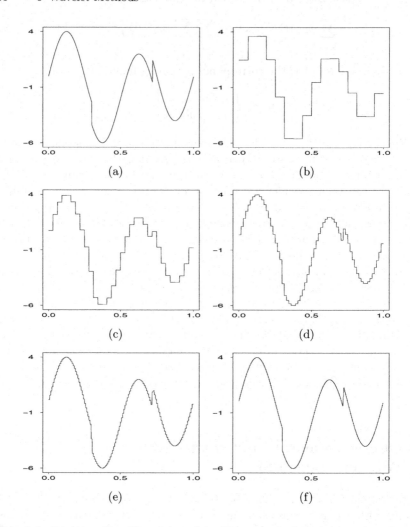

Fig. 5.1. Function HeaviSine (a) and its Haar-wavelet-based coarser versions at resolution levels 3 (b), 4 (c), 5 (d), 6 (e), and 7 (f).

The values of the discretized scaling and wavelet functions computed at $t_i, i = 1, \ldots, n$, will be denoted by ϕ_i and ψ_{lki}. Depending on boundary conditions, the discretization may or may not lead to orthogonal vectors. Here we assume that the necessary boundary corrections are made and $(\phi_1, \ldots, \phi_n)$ and $(\psi_{lk1}, \ldots, \psi_{lkn}), l = 0, \ldots, L - 1, k = 0, \ldots, 2^l - 1$, are orthogonal vectors. Please refer to Cohen et al. (1993) for information on the boundary correction.

In the case of the Haar wavelet, the discretized scaling and wavelet functions are

$$\phi_i = \frac{1}{\sqrt{n}}\phi(t_i),$$

$$\psi_{lki} = \frac{1}{\sqrt{n}}\psi_{lk}(t_i).$$

The discrete wavelet transform (DWT) computed for scale l and translation k will be

$$d_{lk} = \sum_{i=1}^{n} f_i \overline{\psi_{lki}},$$

and the coefficient for the coarse approximation of f at scale 0 will be $c_0 = \sum_{j=1}^{n} f_j \overline{\phi_j}$.

The discrete wavelet expansion of $f_i, i = 1, \ldots, n$, will be

$$f_i = c_0 \phi_i + \sum_{l=0}^{L-1} \sum_{k=0}^{2^l-1} d_{lk} \psi_{lki}. \tag{5.6}$$

A coarse version of f_i at resolution level l_0 will be

$$f_i = c_0 \phi_i + \sum_{l=0}^{l_0} \sum_{k=0}^{2^l-1} d_{lk} \psi_{lki}. \tag{5.7}$$

The higher-resolution wavelet terms can then be included sequentially and will provide a succession of coarser versions of f_i. Analogously to the example provided by Figure 5.1 for the continuous case, a look at the succession of approximations may highlight small-, medium-, and large-scale features of the discretized function.

In practice, the DWT is performed with a fast cascade algorithm (Mallat, 1989) with $O(n)$ operations. For details on the cascade algorithm, please refer to Mallat (1999) and Vidakovic (1999).

5.4.1 Application: Image Compression

We illustrate here the power of the discrete wavelet transform for image compression. Figure 5.2(a) presents a 512 × 512 image of the sunset in Búzios, Rio de Janeiro, Brazil. Using the R package Wavethresh (Nason, 1998), we applied the wavelet transform to the sunset image using the Daubechies wavelets with 6 vanishing moments and periodic boundary treatment. After that, we assigned value zero to the 95% smallest (in absolute value) wavelet coefficients, an operation known in the wavelets literature as thresholding. Finally, we applied the inverse wavelet transform to the thresholded wavelet coefficients to obtain the image shown in Figure 5.2(b). The compression of the image is incredible: with only 5% of the wavelet coefficients, Figure 5.2(b) is indistinguishable from Figure 5.2(a).

(a)   (b)

Fig. 5.2. Sunset in Búzios, Rio de Janeiro, Brazil. Example of image compression with Daubechies wavelets. (a) Original image. (b) Reconstructed image using the 5% largest wavelet coefficients.

5.5 Bayesian Nonparametric Regression with Wavelets

Wavelets have been shown to possess near-minimax optimality when used for nonparametric regression (Donoho and Johnstone, 1994, 1995). The procedure is very simple: the discrete wavelet transform is applied to the observed data, and the resulting empirical wavelet coefficients are shrunk toward zero and transformed back to the data domain by the inverse discrete wavelet transform. Several approaches based on the Bayesian paradigm have been shown to provide a smaller mean square error than the procedures of Donoho and Johnstone (1994, 1995) for different test functions (e.g., Chipman et al., 1997; Clyde et al., 1998). Moreover, these Bayesian approaches may incorporate prior beliefs about the regularity of the function through the joint prior distribution of the wavelet coefficients. The several Bayesian approaches differ with respect to the prior distribution of the coefficients, the prior distribution of the other model parameters, the loss function, and the estimation method.

Initial Bayesian approaches assumed independent prior distributions for each wavelet coefficient with different hyperparameters at different resolutions (Chipman et al., 1997; Abramovich et al., 1998; Vidakovic, 1998; Clyde et al., 1998). Chipman et al. (1997) assumed as independent priors for the coefficients a mixture of two normals, whereas Abramovich et al. (1998) and Clyde et al. (1998) assumed a mixture of a normal and a point mass at zero, and Vidakovic (1998) assumed Student-t priors. With respect to the loss function, Chipman et al. (1997), Vidakovic (1998), and Clyde et al. (1998) considered the quadratic loss, whereas Abramovich et al. (1998) considered the absolute loss function. An interesting aspect of the use of the absolute loss function is that the estimator is the posterior median; as a result, some wavelet coefficients may be estimated as equal to zero. As this may lead to signal representations based on a few wavelet coefficients different from zero, this may be particularly useful when the objective is to compress the signal. An alternative that also leads to wavelet coefficients estimated as zero is to use Bayesian hypothesis tests (e.g., see Vidakovic, 1998). As the specification of the priors for the hyperparameters is not an easy task in wavelet nonparametric regression, Abramovich et al. (1998) and Clyde and George (2000) have proposed empirical Bayes estimation for wavelets. Finally, the model can be made more robust by using Student-t errors (Vidakovic, 1998; Clyde and George, 2000).

Vannucci and Corradi (1999a) have presented a recursive approach to computing the covariances between the empirical wavelet coefficients within and across scales. Based on the empirical wavelet coefficients' covariance structure, Vannucci and Corradi (1999a,b) have proposed a joint prior distribution for the wavelet coefficients. In a related approach, Vidakovic and Müller (1995) have proposed a joint prior distribution for the wavelet coefficients that assumes independence across scales and dependence within scales. In a fairly different approach, Crouse et al. (1998) and Nowak (1999) model the dependence between wavelet coefficients using hidden Markov models on trees. For an introduction to hidden Markov models on trees, see Chapter 8.

5.5.1 Statistical Model

In nonparametric regression with wavelets, it is assumed that observations $y_1, \ldots, y_n$ are taken with measurement error at equally spaced points $t_1, \ldots, t_n$, where, as in Section 5.4, $t_i = (i-1)2^{-L}$. Formally

$$y_i = f(t_i) + \epsilon_i = f_i + \epsilon_i, \tag{5.8}$$

where f_i can be expanded with the discrete wavelet expansion (5.6). Using that expansion, Equation (5.8) can be rewritten in matrix notation as

$$\mathbf{y} = \mathbf{W}\boldsymbol{\beta} + \boldsymbol{\epsilon}, \tag{5.9}$$

where $\mathbf{y} = (y_1, \ldots, y_n)'$, $\boldsymbol{\epsilon} = (\epsilon_1, \ldots, \epsilon_n)' \sim N(\mathbf{0}, \sigma^2 \mathbf{I})$, $\boldsymbol{\beta} = (c_0, d_{00}, d_{10}, d_{11},$
$\ldots, d_{L-1,2^{L-1}-1})'$ is the n-dimensional vector of wavelet coefficients and

$$\mathbf{W} = \begin{bmatrix} \phi_1 & \psi_{001} & \psi_{101} & \psi_{111} & \cdots & \psi_{L-1,2^{L-1}-1,1} \\ \phi_2 & \psi_{002} & \psi_{102} & \psi_{112} & \cdots & \psi_{L-1,2^{L-1}-1,2} \\ \vdots & \vdots & \vdots & \vdots & & \vdots \\ \phi_n & \psi_{00n} & \psi_{10n} & \psi_{11n} & \cdots & \psi_{L-1,2^{L-1}-1,n} \end{bmatrix} \tag{5.10}$$

is the $n \times n$ orthogonal matrix that contains the discretized scaling and wavelet functions.

The discrete wavelet transform of the observations $y_1, \ldots, y_n$ corresponds to the multiplication of $\mathbf{y}$ by $\mathbf{W}'$. Let $\hat{\mathbf{d}} = \mathbf{W}'\mathbf{y}$ be the vector of empirical wavelet coefficients. In addition, let $\boldsymbol{\varepsilon} = W'\boldsymbol{\epsilon} \sim N(\mathbf{0}, \sigma^2 \mathbf{I})$. Then, regression (5.9) becomes

$$\widehat{\mathbf{d}} = \boldsymbol{\beta} + \boldsymbol{\varepsilon}. \tag{5.11}$$

As wavelet representations of functions will typically have a few large (in absolute value) wavelet coefficients and many zero or close to zero, it is reasonable to assume sparsity of the vector $\boldsymbol{\beta}$. This knowledge can be incorporated in the model with independent mixtures of a Gaussian and a point mass at zero as prior distributions for the elements of $\boldsymbol{\beta}$ as proposed by Abramovich et al. (1998) and Clyde et al. (1998),

$$d_{lk} \sim w_l N(0, c_l \sigma^2) + (1 - w_l)\delta(0),$$

where $\delta(0)$ represents a point mass at zero, w_l is the expected fraction of nonzero wavelet coefficients at level l, and c_l is related to the magnitude of the coefficients at level l.

A full Bayesian analysis requires the assignment of priors for the hyperparameters $\sigma^2, w_0, \ldots, w_{L-1}, c_0, \ldots, c_{L-1}$. Moreover, a full Bayesian analysis may require MCMC-based methods. This type of analysis takes into account all sources of uncertainty and can be extended to more complex situations, such as, for example, nonequally spaced regression and hierarchical functional analysis. For a nice introduction to MCMC methods for wavelets, see Müller and Vidakovic (1999b).

In practice, very often a lack of prior knowledge about the function f will prevent the elicitation of priors for the hyperparameters. Moreover, an MCMC-based full Bayesian analysis may be computationally too expensive. As an alternative, empirical Bayes approaches are often used (e.g., Clyde and George, 2000). Typically, an empirical Bayes approach estimates the hyperparameters $\sigma^2, w_0, \ldots, w_{L-1}, c_0, \ldots, c_{L-1}$ using method of moments or maximum likelihood and then replaces the hyperparameters with their estimates in the posterior analysis of the wavelet coefficients.

5.5.2 Estimation of Hyperparameters

An approach that works well in practice is to estimate σ^2 using the median absolute deviation estimator (Donoho and Johnstone, 1995; Donoho et al., 1995)

$$\hat{\sigma} = \frac{1}{0.6745} \text{median}_k(|\hat{d}_{L-1,k}|).$$

This estimator uses the fact that most empirical wavelet coefficients at the finest resolution level will be just white noise. Moreover, use of the median provides a robust estimator in case some signal is present at the finest resolution level through a few large coefficients.

Conditional on $\sigma = \hat{\sigma}$, the remaining hyperparameters can be estimated via maximum likelihood. Integrating out $\boldsymbol{\beta}$, the empirical wavelet coefficients will be conditionally independent given $w_0, \ldots, w_{L-1}, c_0, \ldots, c_{L-1}$. In that case,

$$\hat{d}_{lk} \sim w_l N(0, (1 + c_l)\sigma^2) + (1 - w_l)N(0, \sigma^2),$$

and the likelihood function of w_l, c_l will be proportional to

$$\prod_k \left\{ \frac{w_l}{\sqrt{1 + c_l}} \exp\left[-\frac{1}{2(1 + c_l)\sigma^2} \left(\hat{d}_{lk}^2\right) \right] + (1 - w_l) \exp\left[-\frac{1}{2\sigma^2} \left(\hat{d}_{lk}^2\right) \right] \right\}.$$

In order to find $\hat{w}_0, \ldots, \hat{w}_{L-1}, \hat{c}_0, \ldots, \hat{c}_{L-1}$, the maximization of the likelihood function above has to be performed numerically. Clyde and George (2000) maximize this function with nonlinear Gauss-Seidel iteration, whereas Johnstone and Silverman (1998) use an EM algorithm (Dempster et al., 1977); in practice, both methods have similar performance.

The estimation of one pair (w_l, c_l) for each resolution level may lead to high frequentist variance of hyperparameter estimates and overly adaptive estimated curves. In order to ameliorate the situation, it is advisable to write (w_l, c_l) in terms of a few parameters and then maximize the likelihood function for those parameters. In addition, it is advisable to perform shrinkage or thresholding of wavelet coefficients only at resolution levels higher than a prespecified level l_0. Here we use a parameterization similar to that of Abramovich et al. (1998):

$$c_l = 2^{-\alpha l} C_1,$$
$$w_l = 2^{-\beta l} C_2,$$

$l = l_0, \ldots, L - 1$. Abramovich et al. (1998) discuss the relationship between (α, β) and Besov space parameters, and thus how prior knowledge about the function's regularity can be incorporated in the prior for the wavelet coefficients through (α, β). The hyperparameters C_1 and C_2 can be easily estimated via maximum likelihood.

5.5.3 Empirical Bayes Estimation of Wavelet Coefficients

Once the hyperparameters have been estimated, their estimates are used as true values in the posterior analysis of the wavelet coefficients. Given σ^2, $w_0, \ldots, w_{L-1}, c_0, \ldots, c_{L-1}$, the elements of $\boldsymbol{\beta}$ will be a posteriori conditionally independent. Moreover, the posterior distribution of d_{lk} will be a mixture of a Gaussian and a point mass at zero (e.g., see Clyde and George, 2000),

$$d_{lk}|\mathbf{y} \sim p_{lk} N\left[\frac{c_l}{1 + c_l} \hat{d}_{lk}, \frac{c_l}{1 + c_l}\sigma^2 \right] + (1 - p_{lk})\delta(0), \qquad (5.12)$$

where $p_{lk} = O_{lk}/(1 + O_{lk})$ is the posterior probability that d_{lk} is different from zero and O_{lk} is the posterior odds that d_{lk} is different from zero:

$$O_{lk} = (1 + c_l)^{-0.5} \frac{w_l}{1 - w_l} \exp\left\{0.5\frac{c_l}{1 + c_l}\frac{d_{lk}^2}{\sigma^2}\right\}.$$

Depending on the loss function used, different estimators of d_{lk} can be obtained from the posterior distribution (5.12). Use of square loss leads to the posterior mean (Clyde et al., 1998)

$$E(d_{lk}|\mathbf{y}) = p_{lk}\frac{c_l}{1 + c_l}\hat{d}_{lk}, \qquad (5.13)$$

which is a shrinkage estimator.

Use of the absolute loss function leads to the posterior median (Abramovich et al., 1998; Johnstone and Silverman, 2004, 2005), which is a threshold estimator equal to zero if

$$p_{lk}\Phi\left[\left(\frac{c_l}{1 + c_l}\right)^{0.5}\frac{|\hat{d}_{lk}|}{\sigma}\right] \leq 0.5, \qquad (5.14)$$

where Φ is the standard Gaussian cumulative distribution function. Thus, if $p_{lk} \leq 0.5$, the posterior median is necessarily equal to 0. If inequality (5.14) is not satisfied, then the posterior median is equal to

$$\frac{c_l}{1 + c_l}\hat{d}_{lk} - \text{sign}(\hat{d}_{lk})\sigma\left(\frac{c_l}{1 + c_l}\right)^{0.5}\Phi^{-1}\left(\frac{0.5}{p_{lk}}\right).$$

Another threshold estimator (Clyde and George, 2000) is obtained by using the posterior mean conditional on the most probable component of the mixture in Equation (5.12). Thus, the estimator is equal to zero if $p_{lk} \leq 0.5$ and equal to $c_l(1 + c_l)^{-1}\hat{d}_{lk}$ if $p_{lk} > 0.5$. Clyde and George (2000) report that the performance of this last estimator is worse than the performance of the square-loss-based estimator given in Equation (5.13).

5.5.4 Application

Here we illustrate the use of wavelets for nonparametric regression with an application to the Doppler test function (Donoho and Johnstone, 1994):

$$\sqrt{t(1 - t)} * \sin\left(2\pi\frac{1 + 0.05}{t + 0.05}\right).$$

We rescaled this function and added independent $N(0, 1)$ noise so that the signal-to-noise ratio was equal to 5. Figure 5.3(a) presents the Doppler function, and Figure 5.3(b) presents noisy observations.

We applied Daubechies's least asymmetric wavelet of order 8 (Daubechies, 1992) using Wavethresh (Nason, 1998). Figure 5.4(a) presents the wavelet coefficients by resolution level. The lower resolution levels, those related to the large-scale features, are dominated by some large wavelet coefficients.

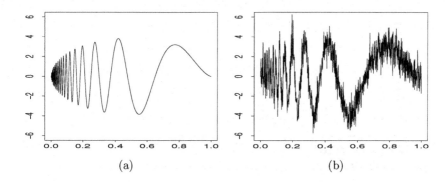

Fig. 5.3. (a) Doppler function. (b) Doppler function with noise. Signal-to-noise ratio equal to 5.

On the contrary, the wavelet coefficients at the higher resolution levels are confounded with noise.

We used the empirical Bayes approach described in this chapter with $l_0 = 3$, $\alpha = 1$, $\beta = 1$, and C_1 and C_2 estimated via maximum likelihood as equal to 1701.38 and 5.90, respectively. These values for α and β assume that both the prior variance of nonzero wavelet coefficients and the prior probability of a nonzero wavelet coefficient decrease 50% from a given resolution level to the next higher resolution level. Figures 5.4(b) and 5.4(c) present the wavelet coefficients' posterior mean and median, respectively. When compared with Figure 5.4(a), these figures display a much clearer picture of the multiresolution behavior of the function to be estimated. The main features at each resolution level are located in the regions where the estimated wavelet coefficients are large. Figures 5.4(b) and 5.4(c) indicate that large-scale features are located close to $t = 1$, and as smaller scales are considered, the important features move closer to $t = 0$.

As the posterior mean is a shrinkage estimator and the posterior median is a thresholding estimator, Figure 5.4(c) displays a much more dramatic reduction in the wavelet coefficients than Figure 5.4(b). The difference between the wavelet coefficients' posterior mean and median carries on to the estimation of the Doppler function. Figures 5.5(a) and 5.5(b) show the estimated Doppler function using the wavelet coefficients' posterior mean and median, respectively. The Doppler function estimate based on the posterior median is smoother and visually much more pleasant.

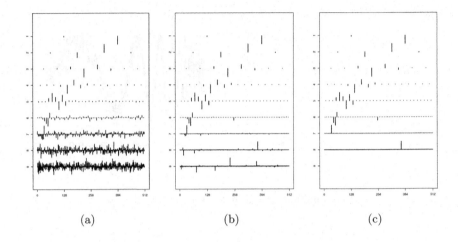

(a)   (b)   (c)

Fig. 5.4. Wavelet coefficients. Resolution increases from top to bottom. Here, the posterior mean is a shrinkage estimator, whereas the posterior median is a threshold estimator. (a) Empirical wavelet coefficient. (b) Posterior mean. (c) Posterior median.

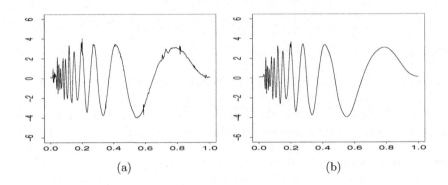

Fig. 5.5. Doppler function estimates based on wavelet coefficients' (a) posterior mean and (b) posterior median.

5.6 Other Statistical Applications of Bayesian Wavelet Analysis

Bayesian wavelet analysis has been successfully applied to a myriad of problems in several different fields. A full treatment of the subject cannot be fit within a single chapter; it would rather need a book by itself. Here we mention just some of the many successful applications of Bayesian wavelet analysis: in time series, estimation of ARFIMA(p,d,q) parameters (Ko and Vannucci, 2006a), detection of multiple change points of a long memory parameter (Ko and Vannucci, 2006b), and estimation of spectral density (Pensky et al., 2006); in bioinformatics, extraction of structural features from proteins (Liò and Vannucci, 2000; Vannucci and Liò, 2001); analysis of turbulence flows (Katul et al., 2001); analysis of network traffic (Kim et al., 2004; Kwon et al., 2006); analysis of functional magnetic resonance imaging data (Sendur et al., 2005); estimation of the link function in single-index models (Park et al., 2005); hierarchical functional analysis (Morris et al., 2003); and spectroscopic calibration (Brown et al., 2001; Vannucci et al., 2003). Please refer to Vannucci (2007) for more details on statistical applications of wavelets.

Part III

Explicit Multiscale Models

6

Overview of Explicit Multiscale Models

Multiscale modeling arises in a wide variety of applications. As discussed in Chapter 1, there are at least three classes of problems that can be modeled most effectively within a multiscale framework. In the first type, data are observed at different spatial scales and the model is used to integrate the information from the different scales. In the second type, data are observed only at the finest scale and the model is used to induce a particular process at that scale. In the third type, the observed data are related nonlocally and nonlinearly to an underlying multiscale process, and the model is used as a prior for that process.

Many multiscale procedures have appeared in the engineering literature and have focused on the development of coarser representations of the phenomenon of interest in order to obtain fast computational algorithms. In most of that literature, the statistical structure is isolated from scale to scale, and thus there is no consistent joint multiresolution statistical model as, for example, in the work of Saquib et al. (1996), Comer and Delp (1999), and Pizurica et al. (2002). An exception is the work developed by Allan S. Willsky and his coauthors in the past decade or so, which we review in Chapter 7 (see also Willsky, 2002). In that body of work, Willsky et al. developed multiscale models in dyadic and quad trees in which sites of a given level are conditionally independent given the immediate coarser level. This allows a state-space representation of those models, and thus a variant of the Kalman filter can be used for inference (for references on state-space models and the Kalman filter, see Harvey, 1989, and West and Harrison, 1997).

Fully Bayesian approaches generalizing that body of work are reasonably new and have been applied to a variety of fields. For example, Kolaczyk (1999) introduces Bayesian multiscale models based on recursive dyadic partitions for Poisson processes, Nowak and Kolaczyk (2000) use such a framework to solve Poisson inverse problems, Nowak (1999) proposes a multiscale hidden Markov model for Bayesian image analysis, Kolaczyk and Huang (2001) construct a multiscale model for spatial aggregation, and Huang et al. (2002) use multiscale models to perform fast spatial prediction for global processes. Although those models lead to very efficient algorithms, they introduce artifacts in the analysis such as the blocky behavior pointed out by Irving et al. (1997). A possible remedy to those artifacts is to define a multiscale model on a more general graph as proposed by Huang and Cressie (2001), where each site at a given resolution level depends on more than one site at the immediate coarser level, and the knowledge of the immediate coarser level decorrelates the sites at that given resolution level. The model by Huang and Cressie (2001) leads to smoother processes than the tree-based models at each resolution level but, like the tree-based models, does not take into account the dynamics at each level of resolution. Some of these models are reviewed in Chapter 9.

An alternative approach that does model the dynamics at each level of resolution is the multiscale random field approach of Ferreira et al. (2005). This class of multiscale random fields is useful for modeling processes that live and possibly can be observed at different levels of resolution. Each level of resolution is connected with the immediately finer level through a linear function plus Gaussian noise. Initially, Markov random field processes are assigned for each level of resolution, and then Jeffrey's rule of conditioning (see, for example, Jeffrey, 1988) is used to revise the implied distributions and ensure that the probability distributions of the different levels are compatible. As these models assume the existence of dependence between sites of a given resolution level even conditional on the immediate coarser level, they do not have the undesirable block effects, and they do incorporate the dynamics at each level. In fact, they are able to accommodate fairly smooth processes at the different levels of resolution. Moreover, this class of models has the ability both to combine information across levels of resolution and to emulate long memory spatial processes. These models are reviewed in Chapter 10.

As time series processes can be seen as random field processes in one dimension, the same type of multiscale model of Ferreira et al. (2005) can be specialized to time series modeling, as in the work of Ferreira et al. (2006). These multiscale time series models are able to consistently model time series at different levels of resolution (e.g., daily or monthly aggregates of financial or meteorological data) and to coherently combine information across time scales. Several issues particular to time series analysis arise, such as the inclusion of seasonality in the model and the capacity to perform predictions. In particular, when compared with multiscale analyses of time series based on wavelet decompositions, the ability to perform predictions is a great advantage of the multiscale models of Ferreira et al. (2006). These models are reviewed in Chapter 11.

A rather different perspective is presented in Chapter 12, where one may expect that the discrete blocks at different scales fail to line up easily across scales. Whereas the other chapters in this part primarily deal with the case where the number of cells at a finer scale is an integer multiple of the number of cells at a coarser scale, so that a coarse cell can be comprised of a fixed number of finer cells, the change of support methods allow more arbitrary movement across scales. We include the change of support modeling chapter in this section because it also describes an explicit model for moving between different resolutions, although the flavor of the models is noticeably different from those in the rest of this section.

While we have tried to include a wide range of broadly applicable methods, we regret that this section is not an exhaustive compilation of Bayesian explicit multiscale methods. For example, a neat approach that has appeared in the context of the analysis of deterministic computer simulators at multiple levels of resolution is that of Kennedy and O'Hagan (2000). Each level is modeled with a Gaussian process, and the levels are linked through an autoregressive process. The setup thus gives a joint multivariate normal distribution for the combined data across all scales, simplifying the posterior analysis.

Before moving on to the methodological chapters, the rest of this chapter presents a brief review of trees and some tree models.

6.1 Tree Definition

A tree is a graph formed by nodes connected by line segments containing no closed loops. Here we consider rooted trees. In a rooted tree, one of the nodes is the root node, and nodes at the same distance from the root node are said to be in the same level of resolution. Levels closer to the root are said to be coarser and levels farther away are said to be finer. Two immediately connected nodes at a coarser and a finer level are said to be the parent and a child of each other, respectively. Trees with two or four children per node are called dyadic trees or quadtrees, respectively. Figure 6.1 represents three levels of a dyadic tree.

In general, a model on a tree with L levels of resolution is considered, where (l, j) denotes the jth node at the lth level, $l = 0, \ldots, L - 1$. It is assumed that node $(l, j), l \geq 1$, has one parent node at the immediate coarser level $l - 1$ and $d_{l,j}$ descendant nodes at the immediate finer level $l + 1$. Let D_{lj} denote the set of nodes of the $(l + 1)$th level that are descendants of (l, j), and let A_{lj} denote the node of the $(l - 1)$th level that is the ancestor of (l, j). Denote by n_l the number of nodes at the lth level; we can recursively compute $n_l = \sum_{j=1}^{n_{l-1}} d_{l-1,j}$ and typically $n_0 = 1$.

6.2 Classification and Regression Trees

The basic tree model in statistics is CART (classification and regression trees), also known as recursive partitioning regression (Breiman et al., 1984). In a regression setting, the idea is to use a tree to partition the space of the explanatory variables, fitting the response with a constant value in each partition. Typically, dyadic trees with binary splits are used, but by allowing recursive splitting, variables may be revisited and thus nonbinary partitions can be achieved. Also, typical splitting rules are purely recursive, so that splits

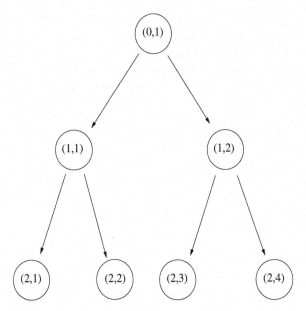

Fig. 6.1. Three levels of a dyadic tree

at one node are independent of any possible splits at another node at the same level. For example, in Figure 6.1, the root node $(0, 1)$ may represent a split on the first explanatory variable, with node $(1, 1)$ representing all cases with $x_1 \leq r_1$ and $(1, 2)$ containing all cases with $x_1 > r_1$. Then the split between nodes $(2, 1)$ and $(2, 2)$ could split again on the first variable, with $(2, 1)$ containing cases with $x_1 \leq r_2 < r_1$, and $(2, 2)$ containing cases with $r_2 < x_1 \leq r_1$. But node $(1, 2)$ can split differently, such as by using a second explanatory variable x_2. Thus the splits are independent conditioned on the parents of the nodes. The fitted value is then the average of the responses of the cases in the partition. For a classification problem, the tree would predict the fitted probability of class membership to be the empirical probability of the cases in that partition (the number of cases in the partition in that class divided by the total number of cases in the partition). More details on tree models can be found in Hastie et al. (2001) and the references therein.

   CART has gained popularity because it is easy to use, produces a model that can often give a meaningful interpretation, and has been found to work reasonably well in a wide variety of problems. Tree models can be extended to allow the fitting of linear regression or other more complex models over each of the regions (Chipman et al., 2002; Gramacy and Lee, 2006). Bayesian versions of CART have also been developed (Denison et al., 1998; Chipman et al., 1998). The key ingredients are the specification of the prior over the space of trees and a mechanism for mixing over this space during MCMC. Care must be taken, as the posterior over the tree space tends to be multimodal.

Partition models can also be generalized beyond trees, such as with a Voronoi tessellation partition (Denison et al., 2002).

In some sense, CART can be viewed as a multiscale model in that each level of the tree represents a successive refinement. In this way, multiscale models can be built somewhat analogously to the convolution and wavelet models in Chapters 4 and 5. However, the basic CART model is somewhat suboptimal for this purpose, as the different branches of the tree are typically unrelated to each other, and so the successive refining happens in a rather nonsystematic way with respect to the multiscale nature of the problem. For this reason, we focus on other varieties of tree models in the following chapters.

7

Gaussian Multiscale Models on Trees

Many multiscale approaches have appeared in the engineering literature, with a focus on the development of coarser representations of the phenomenon of interest in order to obtain fast computational algorithms. With this objective in mind, a particularly effective multiscale framework was developed by Allan S. Willsky and his coauthors in the 1990s (Basseville et al., 1992a,b; Luettgen et al., 1993; Chou et al., 1994a,b; Luettgen et al., 1994; Luettgen and Willsky, 1995a,b; Irving et al., 1997; Frakt and Willsky, 1998; Daoudi et al., 1999). In that work, Willsky et al. developed multiscale models on trees in which nodes of a given level are conditionally independent given the immediate coarser level. Those models admit a state-space representation and, as a result, a variant of the Kalman filter can be used for inference (for references on state-space models and the Kalman filter, see Harvey, 1989, and West and Harrison, 1997). For a detailed review of multiscale modeling on trees from an engineering point of view, the reader is referred to Willsky (2002). In this chapter, we review from a Bayesian statistical point of view the main points of Willsky's multiscale framework (WMF).

In a seminal set of papers, Basseville et al. (1992a,b) introduced the basis for multiscale autoregressive modeling on dyadic trees. Building on those developments, Chou et al. (1994a,b) introduced Gaussian multiscale models on dyadic trees and a corresponding optimal estimation algorithm. In addition, they presented examples of the use of these multiscale models on dyadic trees for the denoising of signals, the incorporation of information from multiple resolutions, and the estimation of motion. Use of the multiscale estimation algorithm of Chou et al. (1994a,b) is not restricted to dyadic trees, and it can be used for Gaussian multiscale models on quadtrees. Moreover, autoregressive multiscale models on quadtrees may be used to approximate Gaussian random fields, as pointed out by Luettgen et al. (1993) with the approximation of Markov random fields. Luettgen et al. (1994) used multiscale dynamic models on quadtrees as priors for the computation of optical flow, a computationally intensive inverse problem. Finally, Luettgen and Willsky (1995a) developed an algorithm for likelihood computation for multiscale models on trees that may be used for maximum likelihood estimation or for the computation of likelihood ratio tests.

One of the problems of models on quadtrees is that the resulting covariance function at the highest resolution level may be very blocky. In order to reduce that problem, Irving et al. (1997) proposed the use of overlapping trees. Another possible solution was proposed by Huang and Cressie (1997), who introduced multiscale dynamic models on acyclic directed graphs and developed a fast optimal estimation algorithm for the state-space process, while Huang and Cressie (2001) applied these models to command-and-control problems.

There are many parallels between multiscale autoregressive models on trees and wavelets. Both consider multiple scales of resolution, and multiscale autoregressive models on trees can mimic the wavelet reconstruction algorithm (Daoudi et al., 1999). In its most basic version, wavelet analysis decomposes a signal in a sequence of details at different levels of resolution in a descriptive manner, and statistical modeling is then used in order to decide if each detail (wavelet coefficient) is significant or not (e.g., see Vidakovic, 1999; Mallat, 1999). In turn, Willsky's multiscale autoregressive models may be seen as stochastic versions of the wavelet reconstruction algorithm, with independent Gaussian details being added as the resolution level increases. Finally, in contrast to wavelets, the WMF is able to accommodate observations at different levels of resolution.

The ability of WMF to accommodate observations at different resolution levels can be used, for example, to integrate atmospheric data. Let us say that a given process of interest such as the amount of rain can be measured with measurement error by different equipment, such as terrestrial stations, radar, and satellites. These measurements provide information about the process of interest at different levels of resolution and may have different spatial coverages. In the WMF, the underlying process of interest would be modeled with a latent process on a tree, where nodes at a given scale would correspond to different aggregation levels of the process. The nodes at the different scales would be connected through a linear model. The measurements at each resolution would be connected to the respective nodes through an observation equation. The algorithms of Section 7.3 would then be used to integrate the measurement information from the different resolution levels and to provide estimates of the process of interest at the several resolution levels.

Figure 7.1 represents three levels of a latent process on a dyadic tree. Here, x_l denotes the vectorized latent process at the lth level of resolution, $l = 0, \ldots, L - 1$, and x_{lj} denotes the value corresponding to the jth node at the lth level. The latent process at the root node is $x_0 = x_{lj}$. When dealing with time series, Willsky uses dyadic trees, and so $d_{l,j} = 2, \forall (l, j)$. When dealing with 2-D random fields, Willsky uses quadtrees, and in that case $d_{l,j} = 4, \forall (l, j)$. In order to simplify notation, define $x_{D_{lj}} = \{x_{l'j'} | (l', j') \in D_{lj}\}$. Finally, denote by y_l the vectorized (potentially) observed process at the lth level of resolution and by y_{lj} the n_y-dimensional (potentially) observed value at the jth node at the lth level. Note that while in some applications there are observations at all levels of resolution, in other applications there are observations at only one level of resolution, or only some elements of $\mathbf{y}_l$ at each level $l = 0, \ldots, L-1$ are observed. In order to differentiate observed nodes from nonobserved nodes, a dummy variable γ_{lj} assumes the value 1 if $\mathbf{y}_{lj}$ was actually observed and 0 otherwise.

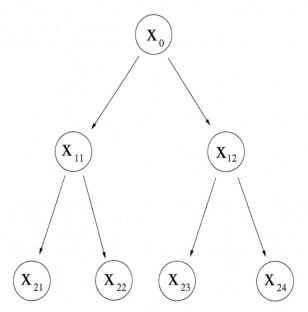

Fig. 7.1. Three levels of a dyadic tree.

7.1 The Model

Assuming that the latent process at nodes of a given level are conditionally independent given the latent process at the immediate coarser level, and assuming that the observations at each node and each level are conditionally independent given the latent process, the joint density function of $(\mathbf{y}_0, \ldots, \mathbf{y}_{L-1}, \mathbf{x}_0, \ldots, \mathbf{x}_{L-1})$ of a generalized version of the WMF model can be written as

$$f(\mathbf{y}_0, \ldots, \mathbf{y}_{L-1}, \mathbf{x}_0, \ldots, \mathbf{x}_{L-1} | \boldsymbol{\theta}) = f(\mathbf{x}_0 | \boldsymbol{\theta}) \left[\prod_{l=1}^{L-1} \prod_{j=1}^{n_l} f(\mathbf{x}_{lj} | \mathbf{x}_{A_{lj}}, \boldsymbol{\theta}) \right]$$

$$\times \left[\prod_{l=0}^{L-1} \prod_{j=1}^{n_l} [f(\mathbf{y}_{lj} | \mathbf{x}_{lj}, \boldsymbol{\theta})]^{\gamma_{lj}} \right], \quad (7.1)$$

where $f(\mathbf{x}_{lj}|\mathbf{x}_{A_{lj}}, \boldsymbol{\theta})$ is the density of $\mathbf{x}_{lj}$ given its ancestral node and $f(\mathbf{y}_{lj}|\mathbf{x}_{lj}, \boldsymbol{\theta})$ is the density of $\mathbf{y}_{lj}$ given the corresponding latent variable $\mathbf{x}_{lj}$. Note that this general framework may include nonnormal observations and nonnormal latent processes. Moreover, $\boldsymbol{\theta}$ is the parameter vector associated with the WMF model of interest and is typically of low dimension. For example, $\boldsymbol{\theta}$ may include the overall mean and a scale parameter.

More specifically, Willsky et al. assume that all the densities above are Gaussian. Under this assumption, the model defined by Equation (7.1) can be rewritten as a Gaussian state-space model on a tree (Chou et al., 1994a,b),

$$\mathbf{y}_{lj} = \mathbf{F}_{lj}\mathbf{x}_{lj} + \mathbf{v}_{lj}, \quad \mathbf{v}_{lj} \sim N(\mathbf{0}, \mathbf{V}_{lj}), \tag{7.2}$$

$$\mathbf{x}_{lj} = \mathbf{G}_{lj}\mathbf{x}_{A_{lj}} + \mathbf{w}_{lj}, \quad \mathbf{w}_{lj} \sim N(\mathbf{0}, \mathbf{W}_{lj}), \tag{7.3}$$

$$\mathbf{x}_0 \sim N(\mathbf{0}, \Sigma_0), \tag{7.4}$$

where $\mathbf{v}_{lj}$ and $\mathbf{w}_{lj}$, $l = 0, \ldots, L-1, j = 1, \ldots, n_l$, are all mutually independent. Moreover, $\mathbf{F}_{lj}$, $\mathbf{G}_{lj}$, $\mathbf{V}_{lj}$, and $\mathbf{W}_{lj}$, $l = 0, \ldots, L-1$, are functions of the parameter vector $\boldsymbol{\theta}$.

Example 7.1. Figure 7.2 presents levels 1, 3, 5, 7, and 9 of a multiscale process on the interval $[0, 1]$. At each level, the process is a step function, and each sub-interval corresponds to a node of an autoregressive process on a dyadic tree. This process was generated with $x_0 \sim N(0, 5)$, $F_{lj} = 1$, $G_{lj} = 1$, $V_{lj} = 1/(l+1)$, and $W_{lj} = 10/2^{l+1}$, $l = 0, \ldots, L-1, j = 1, \ldots, n_l$. The reduction of the variance V_{lj} for finer scales is typical of problems that have information at different levels of resolution, higher-resolution information generally being more precise. The reduction of the variance W_{lj} following a power law leads to a fractal-like process (for a reference on fractal processes, see Mandelbrot, 1999). It is noteworthy that the manner in which the finer levels derive from the coarser levels leads to a long memory dependence process at the finest level of resolution.

7.2 Covariance Structure

The covariance structure of the state-space model on a tree defined by Equations (7.2), (7.3), and (7.4) was derived by Chou et al. (1994a). Note that the tree has only one node at the coarsest level 0. Assume that the corresponding latent value, $\mathbf{x}_0$, is normally distributed with zero mean and covariance matrix $V(\mathbf{x}_0)$. Thus $E[\mathbf{x}_{lj}] = 0, \forall l, j$. Moreover, the covariance matrix $V(\mathbf{x}_{lj}) = E[\mathbf{x}_{lj}\mathbf{x}_{lj}']$ can be recursively computed through the Lyapunov equation

$$V(\mathbf{x}_{lj}) = \mathbf{G}_{lj}V(\mathbf{x}_{A_{lj}})\mathbf{G}_{lj}' + \mathbf{W}_{lj}.$$

Then the covariance matrix $V(\mathbf{y}_{lj})$ can be computed as

$$V(\mathbf{y}_{lj}) = \mathbf{F}_{lj}V(\mathbf{x}_{lj})\mathbf{F}_{lj}' + \mathbf{V}_{lj}.$$

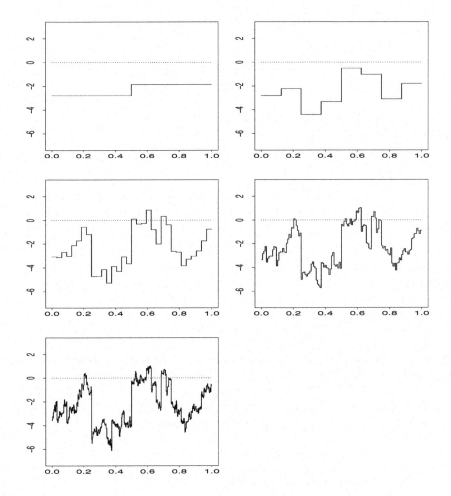

Fig. 7.2. Levels 1, 3, 5, 7, and 9 of a simulated latent process on a dyadic tree, with $x_0 \sim N(0,5)$, $F_{lj} = 1$, $G_{lj} = 1$, $V_{lj} = 1/(l+1)$, and $W_{lj} = 10/2^{l+1}$, $l = 0,\ldots,L-1, j = 1,\ldots,n_l$.

The cross covariance matrix between $\mathbf{x}_{lj}$ and $\mathbf{x}_{l^*j^*}$, denoted by $CV(\mathbf{x}_{lj}, \mathbf{x}_{l^*j^*}) = E[\mathbf{x}_{lj}\mathbf{x}'_{l^*j^*}]$, can be computed as

$$CV(\mathbf{x}_{lj}, \mathbf{x}_{l^*j^*}) = \mathbf{\Phi}_{lj,A(lj,l^*j^*)} V(\mathbf{x}_{A(lj,l^*j^*)}) \mathbf{\Phi}'_{l^*j^*,A(lj,l^*j^*)},$$

where $A(lj, l^*j^*)$ is the most recent common ancestral node of nodes (l, j) and (l^*, j^*), and $\mathbf{\Phi}_{lj,A(lj,l^*j^*)}$ is the transition matrix

$$\mathbf{\Phi}_{lj,A(lj,l^*j^*)} = \begin{cases} I, & (l, j) = A(lj, l^*j^*), \\ \mathbf{G}_{lj}\mathbf{\Phi}_{A_{lj},A(lj,l^*j^*)}, & \text{otherwise.} \end{cases}$$

The cross covariance matrix between $\mathbf{y}_{lj}$ and $\mathbf{y}_{l^*j^*}$ is equal to

$$CV(\mathbf{y}_{lj}, \mathbf{y}_{l^*j^*}) = E(\mathbf{y}_{lj}\mathbf{y}'_{l^*j^*})$$
$$= E(\mathbf{F}_{lj}\mathbf{x}_{lj}\mathbf{x}'_{l^*j^*}\mathbf{F}'_{l^*j^*})$$
$$= \mathbf{F}_{lj} \, CV(\mathbf{x}_{lj}, \mathbf{x}_{l^*j^*}) \, \mathbf{F}'_{l^*j^*}.$$

If $\mathbf{G}_{lj} = \mathbf{G}, \mathbf{W}_{lj} = \mathbf{W}$ and the eigenvalues of $\mathbf{G}$ are all smaller than 1 in absolute value, then existence is guaranteed for the limit

$$\lim_{l \to \infty} V(\mathbf{x}_{lj}) = \mathbf{V}_x.$$

In this case, the matrix $\mathbf{V}_x$ is the stationary covariance matrix of the latent process defined by Equation (7.3), and it can be computed as the solution of the equation

$$\mathbf{V}_x = \mathbf{G}\mathbf{V}_x\mathbf{G}' + \mathbf{W}.$$

In addition, if $V(\mathbf{x}_0) = \mathbf{V}_x$ then $V(\mathbf{x}_{lj}) = \mathbf{V}_x, l = 0, \dots, L - 1, j = 1, \dots, n_l$. Moreover, defining $(l_A, j_A) = A(lj, l^*j^*)$, the cross covariance matrix between $\mathbf{x}_{lj}$ and $\mathbf{x}_{l^*j^*}$ will be

$$CV(\mathbf{x}_{lj}, \mathbf{x}_{l^*j^*}) = \mathbf{G}^{l-l_A}\mathbf{V}_x(\mathbf{G}^{l^*-l_A})'$$

Note that the cross covariance matrix between two nodes depends only on their distances from their most recent common ancestral node.

Moreover, if $\mathbf{F}_{lj} = \mathbf{F}$ and $\mathbf{V}_{lj} = \mathbf{V}, l = 0, \dots, L - 1$, then the entire model defined by Equations (7.2) and (7.3) will be stationary. In this case, the stationary covariance matrix of the observations will be

$$\mathbf{V}_y = \mathbf{F}\mathbf{V}_x\mathbf{F}' + \mathbf{V}$$

and the cross covariance matrices will be

$$CV(\mathbf{y}_{lj}, \mathbf{y}_{l^*j^*}) = \mathbf{F} \, CV(\mathbf{x}_{lj}, \mathbf{x}_{l^*j^*}) \, \mathbf{F}'.$$

Example 7.2. Figure 7.3 presents the correlation function between the univariate latent variables at the finest resolution level for a stationary process on a dyadic tree with $G = 0.9$. It is easy to show that in this case the correlation function between x_{lj} and $x_{l^*j^*}$ is $G^{l+l^*-2l_A}$, where l_A is the resolution level of the most recent common ancestral node of nodes (l, j) and (l^*, j^*). Thus, the correlation function is a step function with the ratio between successive steps equal to G^2. Moreover, the length of each step increases by a factor of 2 at each new step. As a consequence, the correlation function decays very slowly.

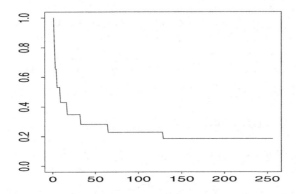

Fig. 7.3. Correlation function between the latent variables at the finest resolution level for a stationary process on a dyadic tree. $G = 0.9$.

7.3 Estimation When θ Is Known

If the parameter vector θ is known, the latent process x can be estimated very quickly through a multiscale estimation algorithm analogous to the Kalman filter. This multiscale estimation algorithm was introduced by Chou, Willsky and Benveniste (1994a) and was explained using Bayesian reasoning by Huang and Cressie (1997). From now on, we will refer to this algorithm as the CWB algorithm. The major difference between the CWB algorithm and the Kalman filter for time series is that while the Kalman filter has a forward in time filtering sweep followed by a backward smoothing sweep, the CWB algorithm has a fine to coarse filtering sweep followed by a coarse to fine smoothing sweep. For information about the Kalman filter, see West and Harrison (1997) and Harvey (1989). For simplicity of exposition, in this section we omit the dependence of the distributions on θ.

The available information set below node (l, j) (that is, the set of observations at descendant nodes of node (l, j)) is simply

$$Y_{lj}^{+} = \{\mathbf{y}_{l'j'} | (l', j') \text{ is a descendant of } (l, j) \text{ and } \gamma_{l'j'} = 1\},$$

while the information set up to node (l, j) is $Y_{lj} = Y_{lj}^{+} \bigcup \{\mathbf{y}_{lj}\}$ if $\gamma_{lj} = 1$ and $Y_{lj} = Y_{lj}^{+}$ if $\gamma_{lj} = 0$. Note that $Y_{lj}^{+} = \bigcup_{(l',j') \in D_{lj}} Y_{l'j'}$ and Y_0 is the set of all observations.

Of particular interest are two distributions of the latent process at node (l, j), each of them conditional on a different set of information: the filtered distribution is conditional on the information up to (l, j), Y_{lj}, and the smoothed distribution is conditional on all the information, Y_0. First the filtered distributions for the latent process at all nodes of the tree are computed in the fine to coarse sweep of the CWB algorithm, and after that the smoothed distributions are computed in the coarse to fine sweep of the CWB algorithm.

With the smoothed distributions, estimators of the latent process and their respective measures of uncertainty can be easily computed. In particular, if the square loss function is used, then the Bayes estimator of $\mathbf{x}_{lj}$ is the posterior mean $E[\mathbf{x}_{lj}|Y_0]$ and the expected loss is the posterior variance $V[\mathbf{x}_{lj}|Y_0]$.

Section 7.3.1 describes the fine to coarse filter of the CWB algorithm and Section 7.3.2 describes the coarse to fine smoother.

7.3.1 Fine to Coarse Filter

The multiscale construction of the latent process x defines a marginal prior for each $\mathbf{x}_{lj}$,

$$\mathbf{x}_{lj} \sim N(\mathbf{0}, \boldsymbol{\Sigma}_{lj}),$$

where $\boldsymbol{\Sigma}_{lj} = \mathbf{G}_{lj}\boldsymbol{\Sigma}_{A_{lj}}\mathbf{G}'_{lj} + \mathbf{W}_{lj}$. Moreover, the coarse to fine construction of the multiscale latent process can be reversed using Bayes' Theorem. More specifically,

$$p(\mathbf{x}_{lj}|\mathbf{x}_{l'j'}) \propto p(\mathbf{x}_{l'j'}|\mathbf{x}_{lj})p(\mathbf{x}_{lj})$$

$$\propto \exp\left\{-\frac{1}{2}(\mathbf{x}_{l'j'} - \mathbf{G}_{l'j'}\mathbf{x}_{lj})'\mathbf{W}_{l'j'}^{-1}(\mathbf{x}_{l'j'} - \mathbf{G}_{l'j'}\mathbf{x}_{lj})\right\}$$

$$\times \exp\left\{-\frac{1}{2}\mathbf{x}'_{lj}\boldsymbol{\Sigma}_{lj}^{-1}\mathbf{x}_{lj}\right\}.$$

Therefore,

$$\mathbf{x}_{lj}|\mathbf{x}_{l'j'} \sim N(\mathbf{B}_{l'j'}\mathbf{x}_{l'j'}, \mathbf{U}_{l'j'}), \quad (l', j') \in D_{lj},$$

where

$$\mathbf{U}_{l'j'} = \left[\boldsymbol{\Sigma}_{lj}^{-1} + \mathbf{G}'_{l'j'}\mathbf{W}_{l'j'}^{-1}\mathbf{G}_{l'j'}\right]^{-1},$$

$$\mathbf{B}_{l'j'} = \mathbf{U}_{l'j'}\mathbf{G}'_{l'j'}\mathbf{W}_{l'j'}^{-1}.$$

The following theorem provides the fine to coarse filter.

Theorem 7.1. *In the Gaussian multiscale model defined by Equations (7.2), (7.3), and (7.4), filtered distributions are given as follows:*

(a) Posterior at the finest resolution level:

$$\mathbf{x}_{L-1,j}|Y_{L-1,j} \sim N(\mathbf{m}_{L-1,j}, \mathbf{C}_{L-1,j}),$$

where

$$\mathbf{m}_{L-1,j} = \gamma_{L-1,j}\boldsymbol{\Sigma}_{L-1,j}\mathbf{F}'_{L-1,j}\left(\mathbf{F}_{L-1,j}\boldsymbol{\Sigma}_{L-1,j}\mathbf{F}'_{L-1,j} + \mathbf{V}_{L-1,j}\right)^{-1}\mathbf{y}_{L-1,j},$$

$$\mathbf{C}_{L-1,j} = \boldsymbol{\Sigma}_{L-1,j} - \gamma_{L-1,j}\boldsymbol{\Sigma}_{L-1,j}\mathbf{F}'_{L-1,j}$$

$$\times \left(\mathbf{F}_{L-1,j}\boldsymbol{\Sigma}_{L-1,j}\mathbf{F}'_{L-1,j} + \mathbf{V}_{L-1,j}\right)^{-1}\mathbf{F}_{L-1,j}\boldsymbol{\Sigma}_{L-1,j}.$$

(b) *Posterior for* $(l', j') \in D_{lj}$: *For some mean* $\mathbf{m}_{l'j'}$ *and covariance matrix* $\mathbf{C}_{l'j'}$,

$$\mathbf{x}_{l'j'} | Y_{l'j'} \sim N(\mathbf{m}_{l'j'}, \mathbf{C}_{l'j'}).$$

(c) *Prior at* (l, j):

$$\mathbf{x}_{lj} | Y_{lj}^+ \sim N(\mathbf{a}_{lj}, \mathbf{R}_{lj}),$$

where

$$\mathbf{R}_{lj} = \left\{ \boldsymbol{\Sigma}_{lj}^{-1} + \sum_{(l', j') \in D_{lj}} \left[(\mathbf{U}_{l'j'} + \mathbf{B}_{l'j'} \mathbf{C}_{l'j'} \mathbf{B}_{l'j'}')^{-1} - \boldsymbol{\Sigma}_{lj}^{-1} \right] \right\}^{-1},$$

$$\mathbf{a}_{lj} = \mathbf{R}_{lj} \left\{ \sum_{(l', j') \in D_{lj}} (\mathbf{U}_{l'j'} + \mathbf{B}_{l'j'} \mathbf{C}_{l'j'} \mathbf{B}_{l'j'}')^{-1} \mathbf{B}_{l'j'} \mathbf{m}_{l'j'} \right\}.$$

(d) *Forecast at* (l, j):

$$\mathbf{y}_{lj} | Y_{lj}^+ \sim N(\mathbf{f}_{lj}, \mathbf{Q}_{lj}),$$

where

$$\mathbf{f}_{lj} = \mathbf{F}_{lj} \mathbf{a}_{lj},$$
$$\mathbf{Q}_{lj} = \mathbf{F}_{lj} \mathbf{R}_{lj} \mathbf{F}_{lj}' + \mathbf{V}_{lj}.$$

(e) *Posterior at* (l, j):

$$\mathbf{x}_{lj} | Y_{lj} \sim N(\mathbf{m}_{lj}, \mathbf{C}_{lj}),$$

with

$$\mathbf{m}_{lj} = \mathbf{a}_{lj} + \gamma_{lj} \mathbf{A}_{lj} \mathbf{e}_{lj},$$
$$\mathbf{C}_{lj} = \mathbf{R}_{lj} - \gamma_{lj} \mathbf{A}_{lj} \mathbf{Q}_{lj} \mathbf{A}_{lj}',$$

where $\mathbf{A}_{lj} = \mathbf{R}_{lj} \mathbf{F}_{lj}' \mathbf{Q}_{lj}^{-1}$ *and* $\mathbf{e}_{lj} = \mathbf{y}_{lj} - \mathbf{f}_{lj}$.

Proof.
The proof is by induction and uses multivariate normal distribution theory and Bayes' Theorem.

(a) At the finest resolution level, $Y_{L-1,j}^+ = \emptyset, j = 1, \ldots, n_{L-1}$, and so $Y_{L-1,j} = \{ \mathbf{y}_{L-1,j} \}$ when $\gamma_{L-1,j} = 1$ and $Y_{L-1,j} = \emptyset$ when $\gamma_{L-1,j} = 0$. As a consequence, if $\gamma_{L-1,j} = 0$ then $p(\mathbf{x}_{L-1,j} | Y_{L-1,j}) = p(\mathbf{x}_{L-1,j})$, implying $\mathbf{m}_{L-1,j} = 0$ and $\mathbf{C}_{L-1,j} = \boldsymbol{\Sigma}_{L-1,j}$. Conversely, when $\gamma_{L-1,j} = 1$ then by Bayes' Theorem

$$p(\mathbf{x}_{L-1,j} | Y_{L-1,j}) = p(\mathbf{x}_{L-1,j} | \mathbf{y}_{L-1,j})$$
$$\propto p(\mathbf{y}_{L-1,j} | \mathbf{x}_{L-1,j}) p(\mathbf{x}_{L-1,j})$$
$$\propto \exp \left\{ -\frac{1}{2} (\mathbf{y}_{L-1,j} - \mathbf{F}_{L-1,j} \mathbf{x}_{L-1,j})' \mathbf{V}_{L-1,j}^{-1} (\mathbf{y}_{L-1,j} \right.$$

$$-\mathbf{F}_{L-1,j}\mathbf{x}_{L-1,j}\Big)\Big\}\exp\left\{-\frac{1}{2}\mathbf{x}'_{L-1,j}\mathbf{\Sigma}^{-1}_{L-1,j}\mathbf{x}_{L-1,j}\right\}$$

$$\propto\exp\left\{-\frac{1}{2}\left(\mathbf{x}_{L-1,j}-\mathbf{m}_{L-1,j}\right)'\mathbf{C}^{-1}_{L-1,j}\right.$$

$$\left.\left(\mathbf{x}_{L-1,j}-\mathbf{m}_{L-1,j}\right)\right\},$$

where $\mathbf{C}^{-1}_{L-1,j} = \mathbf{F}'_{L-1,j}\mathbf{V}^{-1}_{L-1,j}\mathbf{F}_{L-1,j} + \mathbf{\Sigma}^{-1}_{L-1,j}$ and $\mathbf{m}_{L-1,j} = \mathbf{C}_{L-1,j}$ $\mathbf{F}'_{L-1,j}\mathbf{V}^{-1}_{L-1,j}\mathbf{y}_{L-1,j}$. Simple linear algebra shows that these expressions are equivalent to

$$\mathbf{m}_{L-1,j} = \mathbf{\Sigma}_{L-1,j}\mathbf{F}'_{L-1,j}\left(\mathbf{F}_{L-1,j}\mathbf{\Sigma}_{L-1,j}\mathbf{F}'_{L-1,j} + \mathbf{V}_{L-1,j}\right)^{-1}\mathbf{y}_{L-1,j},$$

$$\mathbf{C}_{L-1,j} = \mathbf{\Sigma}_{L-1,j} - \mathbf{\Sigma}_{L-1,j}\mathbf{F}'_{L-1,j}\left(\mathbf{F}_{L-1,j}\mathbf{\Sigma}_{L-1,j}\mathbf{F}'_{L-1,j} + \mathbf{V}_{L-1,j}\right)^{-1}$$
$$\mathbf{F}_{L-1,j}\mathbf{\Sigma}_{L-1,j}.$$

Therefore, $\mathbf{x}_{L-1,j}|Y_{L-1,j} \sim N(\mathbf{m}_{L-1,j}, \mathbf{C}_{L-1,j})$.

(b) It follows directly from the fact that all distributions are Gaussian.

(c) Note that $p(\mathbf{x}_{lj}|Y_{l'j'}) = \int p(\mathbf{x}_{lj}|\mathbf{x}_{l'j'})p(\mathbf{x}_{l'j'}|Y_{l'j'})d\mathbf{x}_{l'j'}$. Thus, $\mathbf{x}_{lj}|Y_{l'j'} \sim N(\mathbf{B}_{l'j'}\mathbf{m}_{l'j'}, \mathbf{U}_{l'j'} + \mathbf{B}_{l'j'}\mathbf{C}_{l'j'}\mathbf{B}'_{l'j'})$. Moreover,

$$p(\mathbf{x}_{lj}|Y^+_{lj}) \propto p(\mathbf{x}_{lj})p(Y^+_{lj}|\mathbf{x}_{lj})$$

$$= p(\mathbf{x}_{lj})\prod_{(l',j')\in D_{lj}}p(Y_{l'j'}|\mathbf{x}_{lj})$$

$$\propto p(\mathbf{x}_{lj})\prod_{(l',j')\in D_{lj}}\frac{p(\mathbf{x}_{lj}|Y_{l'j'})}{p(\mathbf{x}_{lj})}.$$

Therefore, $\mathbf{x}_{lj}|Y^+_{lj} \sim N(\mathbf{a}_{lj}, \mathbf{R}_{lj})$.

(d) It follows directly from (c) and the fact that $\mathbf{y}_{lj}|\mathbf{x}_{lj} \sim N(\mathbf{F}_{lj}\mathbf{x}_{lj}, \mathbf{V}_{lj})$.

(e) Note that $Y_{lj} = Y^+_{lj}\bigcup\{\mathbf{y}_{lj}\}$ when $\gamma_{lj} = 1$ and $Y_{lj} = Y^+_{lj}$ when $\gamma_{lj} = 0$. As a consequence, if $\gamma_{lj} = 0$ then $p(\mathbf{x}_{lj}|Y_{lj}) = p(\mathbf{x}_{lj}|Y^+_{lj})$, implying $\mathbf{m}_{lj} = \mathbf{a}_{lj}$ and $\mathbf{C}_{lj} = \mathbf{R}_{lj}$. Conversely, when $\gamma_{lj} = 1$, then by Bayes' Theorem

$$p(\mathbf{x}_{lj}|Y_{lj}) \propto p(\mathbf{y}_{lj}|\mathbf{x}_{lj})p(\mathbf{x}_{lj}|Y^+_{lj})$$

$$\propto \exp\left\{-\frac{1}{2}\left(\mathbf{y}_{lj} - \mathbf{F}_{lj}\mathbf{x}_{lj}\right)'\mathbf{V}^{-1}_{lj}\left(\mathbf{y}_{lj}\right.\right.$$

$$\left.\left.-\mathbf{F}_{lj}\mathbf{x}_{lj}\right)\right\}\exp\left\{-\frac{1}{2}(\mathbf{x}_{lj} - \mathbf{a}_{lj})'\mathbf{R}^{-1}_{lj}(\mathbf{x}_{lj} - \mathbf{a}_{lj})\right\}$$

$$\propto \exp\left\{-\frac{1}{2}\left(\mathbf{x}_{lj} - \mathbf{m}_{lj}\right)'\mathbf{C}^{-1}_{lj}(\mathbf{x}_{lj} - \mathbf{m}_{lj})\right\},$$

where $\mathbf{C}^{-1}_{lj} = \mathbf{F}'_{lj}\mathbf{V}^{-1}_{lj}\mathbf{F}_{lj} + \mathbf{R}^{-1}_{lj}$ and $\mathbf{m}_{lj} = \mathbf{C}_{lj}\left(\mathbf{F}'_{lj}\mathbf{V}^{-1}_{lj}\mathbf{y}_{lj} + \mathbf{R}^{-1}_{lj}\mathbf{a}_{lj}\right)$. Simple linear algebra shows that these expressions are equivalent to $\mathbf{m}_{lj} =$

$\mathbf{a}_{lj} + \mathbf{A}_{lj}\mathbf{e}_{lj}$ and $\mathbf{C}_{lj} = \mathbf{R}_{lj} - \mathbf{A}_{lj}\mathbf{Q}_{lj}\mathbf{A}'_{lj}$, where $\mathbf{A}_{lj} = \mathbf{R}_{lj}\mathbf{F}'_{lj}\mathbf{Q}_{lj}^{-1}$ and $\mathbf{e}_{lj} = \mathbf{y}_{lj} - \mathbf{f}_{lj}$. Therefore, $\mathbf{x}_{lj}|Y_{lj} \sim N(\mathbf{m}_{lj}, \mathbf{C}_{lj})$.

$\square$

7.3.2 Coarse to Fine Smoother

After the fine to coarse filtering sweep, the analyst will have at the root of the tree the distribution $\mathbf{x}_0|Y_0$, that is, the distribution of the latent process at the root node given all the information. In order to obtain the smoothed distributions $\mathbf{x}_{lj}|Y_0, l = 1, \ldots, L - 1, j = 1, \ldots, n_l$, it will be necessary to send the information Y_0 down the tree. Define $Y_{lj}^- = Y_0 \setminus Y_{lj}$. The following theorem provides the coarse to fine smoother.

Theorem 7.2. *In the Gaussian multiscale model defined by Equations (7.2), (7.3), and (7.4), smoothed distributions are given as follows:*

(a) Posterior for node A_{lj}: For some mean $\mathbf{s}_{A_{lj}}$ and covariance matrix $\mathbf{S}_{A_{lj}}$,

$$\mathbf{x}_{A_{lj}}|Y_0 \sim N(\mathbf{s}_{A_{lj}}, \mathbf{S}_{A_{lj}}).$$

(b) $\mathbf{x}_{lj}|\mathbf{x}_{A_{lj}}, Y_{lj}^- \sim N(\mathbf{k}_{lj}, \mathbf{K}_{lj})$, where

$$\mathbf{K}_{lj} = \left[\mathbf{C}_{lj}^{-1} + \mathbf{W}_{lj}^{-1} - \mathbf{\Sigma}_{lj}^{-1}\right]^{-1},$$

$$\mathbf{k}_{lj} = \mathbf{K}_{lj}\left[\mathbf{C}_{lj}^{-1}\mathbf{m}_{lj} + \mathbf{W}_{lj}^{-1}\mathbf{G}_{lj}\mathbf{x}_{A_{lj}}\right].$$

(c) Posterior for node (l, j):

$$\mathbf{x}_{lj}|Y_0 \sim N(\mathbf{s}_{lj}, \mathbf{S}_{lj}),$$

where

$$\mathbf{S}_{lj} = \mathbf{K}_{lj} + \mathbf{K}_{lj}\mathbf{W}_{lj}^{-1}\mathbf{G}_{lj}\mathbf{S}_{A_{lj}}\mathbf{G}'_{lj}\mathbf{W}_{lj}^{-1}\mathbf{K}_{lj},$$

$$\mathbf{s}_{lj} = \mathbf{K}_{lj}\left[\mathbf{C}_{lj}^{-1}\mathbf{m}_{lj} + \mathbf{W}_{lj}^{-1}\mathbf{G}_{lj}\mathbf{s}_{A_{lj}}\right].$$

Proof.
   Again, the proof is by induction and uses multivariate normal distribution theory and Bayes' Theorem.

(a) It follows directly from the fact that all distributions are Gaussian.
(b) From Bayes' Theorem

$$p(\mathbf{x}_{lj}|\mathbf{x}_{A_{lj}},Y_{lj}) \propto p(\mathbf{x}_{lj}|Y_{lj})p(\mathbf{x}_{A_{lj}}|\mathbf{x}_{lj},Y_{lj})$$
$$= p(\mathbf{x}_{lj}|Y_{lj})p(\mathbf{x}_{A_{lj}}|\mathbf{x}_{lj})$$
$$= p(\mathbf{x}_{lj}|Y_{lj})\frac{p(\mathbf{x}_{lj}|\mathbf{x}_{A_{lj}})p(\mathbf{x}_{A_{lj}})}{p(\mathbf{x}_{lj})}$$
$$\propto \frac{p(\mathbf{x}_{lj}|Y_{lj})p(\mathbf{x}_{lj}|\mathbf{x}_{A_{lj}})}{p(\mathbf{x}_{lj})}$$
$$= \exp\left\{-0.5(\mathbf{x}_{lj}-\mathbf{m}_{lj})'\mathbf{C}_{lj}^{-1}(\mathbf{x}_{lj}-\mathbf{m}_{lj})\right\}$$
$$\times \frac{\exp\left\{-0.5(\mathbf{x}_{lj}-\mathbf{G}_{lj}\mathbf{x}_{A_{lj}})'\mathbf{W}_{lj}^{-1}(\mathbf{x}_{lj}-\mathbf{G}_{lj}\mathbf{x}_{A_{lj}})\right\}}{\exp\left\{-0.5\mathbf{x}_{lj}'\mathbf{\Sigma}_{lj}^{-1}\mathbf{x}_{lj}\right\}}$$
$$\propto \exp\left\{-0.5\left[\mathbf{x}_{lj}'(\mathbf{C}_{lj}^{-1}+\mathbf{W}_{lj}^{-1}-\mathbf{\Sigma}_{lj}^{-1})\mathbf{x}_{lj}\right.\right.$$
$$\left.\left.-2\mathbf{x}_{lj}'(\mathbf{C}_{lj}^{-1}\mathbf{m}_{lj}+\mathbf{W}_{lj}^{-1}\mathbf{G}_{lj}\mathbf{x}_{A_{lj}})\right]\right\}.$$

Therefore, $\mathbf{x}_{lj}|\mathbf{x}_{A_{lj}},Y_{lj} \sim N(\mathbf{k}_{lj},\mathbf{K}_{lj})$, where

$$\mathbf{K}_{lj} = \left[\mathbf{C}_{lj}^{-1}+\mathbf{W}_{lj}^{-1}-\mathbf{\Sigma}_{lj}^{-1}\right]^{-1},$$
$$\mathbf{k}_{lj} = \mathbf{K}_{lj}\left[\mathbf{C}_{lj}^{-1}\mathbf{m}_{lj}+\mathbf{W}_{lj}^{-1}\mathbf{G}_{lj}\mathbf{x}_{A_{lj}}\right].$$

(c) Using $Y_0 = Y_{lj}^{-}\bigcup Y_{lj}$ and $p(\mathbf{x}_{lj}|\mathbf{x}_{A_{lj}},Y_{lj}^{-},Y_{lj}) = p(\mathbf{x}_{lj}|\mathbf{x}_{A_{lj}},Y_{lj})$,

$$p(\mathbf{x}_{lj}|Y_0) = \int p(\mathbf{x}_{lj},\mathbf{x}_{A_{lj}}|Y_0)d\mathbf{x}_{A_{lj}}$$
$$= \int p(\mathbf{x}_{lj}|\mathbf{x}_{A_{lj}},Y_{lj})p(\mathbf{x}_{A_{lj}}|Y_0)d\mathbf{x}_{A_{lj}}.$$

As $\mathbf{x}_{lj}|\mathbf{x}_{A_{lj}},Y_{lj} \sim N(\mathbf{k}_{lj},\mathbf{K}_{lj})$, and $\mathbf{x}_{A_{lj}}|Y_0 \sim N(\mathbf{s}_{A_{lj}},\mathbf{S}_{A_{lj}})$, the result follows from the convolution of Gaussian random vectors.

$\square$

Example 7.3. Figure 7.4 presents the latent process and observations at levels 1, 3, 5, 7, and 9 of the simulated multiscale process of Example 7.1. At each level of resolution, observations are made at the midpoint of some subintervals. Note that not all subintervals have observations, as the probability of the existence of an observation at a particular subinterval at level l is 0.8^l. Thus, in the simulated example, there is an observation at the root node with certainty, while at the finest level the probability of existence of an observation at a particular node is only about 0.13. This is typical of problems where data are available at different scales of resolution, with higher-resolution data being more sparse.

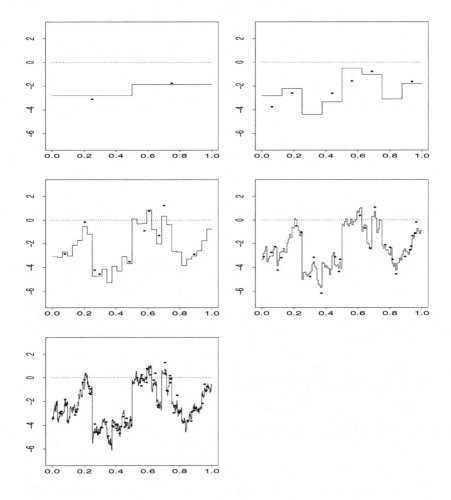

Fig. 7.4. Levels 1, 3, 5, 7, and 9 of a simulated multiscale process on a tree: latent process (solid line) and observations (bullets). $x_0 \sim N(0, 10)$, $F_{lj} = 1$, $G_{lj} = 1$, $V_{lj} = 1/(l + 1)$, and $W_{lj} = 10 \times 2^{l+1}$, $l = 0, \ldots, L - 1, j = 1, \ldots, n_l$.

Figure 7.5, like Figure 7.4, presents the latent process and observations at levels 1, 3, 5, 7, and 9, but in addition presents the smoothed posterior mean and 95% credible intervals. It is remarkable how the estimation methodology recovers most of the features of the latent process at the different levels of resolution. This remarkable performance is the result of the transfer of information through all the levels of resolution, with observations in one level helping to estimate the latent process at the other levels.

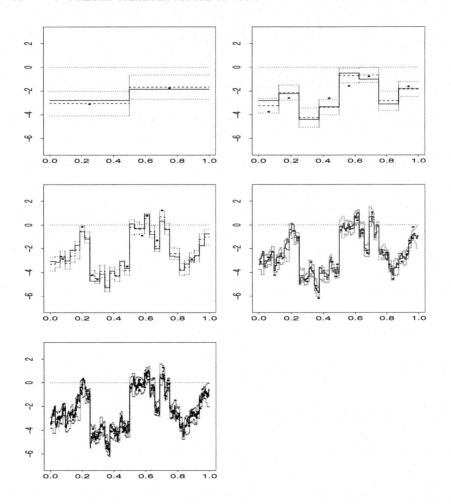

Fig. 7.5. Levels 1, 3, 5, 7, and 9 of a simulated multiscale process on a tree: latent process (solid line), observations (bullets), smoothed means (dashed line), and 95% credible intervals (dotted lines).

7.4 Estimation When θ Is Unknown

If the parameter vector θ is unknown, the estimation algorithm of Section 7.3 cannot be used directly. Instead, the fine to coarse filtering sweep and the coarse to fine smoothing sweep may be integrated within algorithms for the Bayesian estimation of the parameter vector θ and the latent process $\mathbf{x}$. The solution can be a full Bayesian approach implemented with a Markov chain Monte Carlo (MCMC) algorithm (for references on MCMC algorithms, see Gamerman and Lopes, 2006; Robert and Casella, 2005) or an empirical Bayes approach.

A full Bayesian approach requires the elicitation of a prior distribution for θ. As θ is related to the smoothness of the multiscale process, prior knowledge about it can be incorporated in the prior for θ. Moreover, the prior for θ is specific for each particular multiscale model. The full Bayesian approach can be implemented with an MCMC algorithm. The MCMC algorithm simulates a Markov chain whose limiting distribution is the joint posterior distribution of $(\theta, \mathbf{x})$. In the analysis of Gaussian multiscale models on trees, each iteration of the MCMC algorithm can be divided into two steps. In the first step, the latent process is simulated from its full conditional distribution; that is, its posterior distribution conditional on θ. This step uses the fine to coarse filtering sweep of Section 7.3.1 and a coarse to fine sampling sweep that recursively simulates from $\mathbf{x}_{lj}|\mathbf{x}_{A_{lj}}, Y_{lj}$. This step can be seen as a generalization to tree topologies of the forward filtering backward sampling algorithm for state-space models (Carter and Kohn, 1994; Frühwirth-Schnatter, 1994). The second step is constructed according to the multiscale model and the elements of the parameter vector θ. If the full conditional distribution is available for sampling, then this step consists of the simulation of θ from that distribution. Conversely, if the full conditional distribution is not available for sampling, then θ may be simulated with Metropolis or Metropolis-Hastings steps.

Very often in practice, a full Bayesian approach for a Gaussian multiscale model on a tree is not feasible because an MCMC implementation would be too expensive computationally or because a prior for θ would be difficult to elicit. In those cases, an empirical Bayes approach might provide a good solution. Typically, an empirical Bayes approach estimates θ using method of moments or maximum likelihood. Maximum likelihood estimation of θ can be performed by direct maximization of the marginal likelihood or by the EM algorithm. When direct maximization is used, the marginal likelihood of θ can be computed with the fine to coarse filter of Section 7.3.1. Then, the resulting estimate of θ substitutes the true value of θ in the estimation procedures of Sections 7.3.1 and 7.3.2. Alternatively, when the EM algorithm is used, θ and $\mathbf{x}$ are estimated iteratively with two steps at each iteration, the E and the M steps (for references on the EM algorithm, see Dempster et al., 1977; Tanner, 1993). Conditional on the value of θ from the previous iteration, the E step uses the multiscale algorithm of Section 7.3 in order to compute the posterior mean of the latent process $\mathbf{x}$ and assigns this value to $\mathbf{x}$. Conditional on this value of $\mathbf{x}$, the M step consists of finding $\hat{\theta}$ that maximizes the likelihood function (7.1). Even though numerical methods will generally be necessary to perform this maximization step, the computation of the likelihood function is fairly fast as a result of the tree topology associated with the multiscale model.

A good example of the use of an empirical Bayes approach in a multiscale analysis is given in Section 5.5 in the context of wavelet nonparametric regression.

8

Hidden Markov Models on Trees

We present in this chapter hidden Markov models on trees. These models are generalizations of the more traditional one-dimensional hidden Markov models. Traditional hidden Markov models (HMMs) assume hidden states that can take discrete values and are connected through a one-dimensional Markov chain (for a good review on HMMs, see Scott, 2002). In these HMMs, the observations may be discrete or continuous and are conditionally independent given the hidden states. Analogously, hidden Markov models on trees (HMMTs) assume that the values of the latent label process at nodes of a given level are conditionally independent given the latent label process at the immediate coarser level. Moreover, HMMTs assume that the latent label process evolves on a tree in a construction analogous to that described in Chapter 7 for Gaussian processes on trees. More specifically, nodes of a given level are conditionally independent given the immediate coarser level. This conditional independence leads to very fast algorithms for the analysis of HMMTs. HMMTs and variants have been successfully used for image classification (Kato et al., 1996a,b; Laferté et al., 2000) and image segmentation (Bouman and Shapiro, 1994; Comer and Delp, 1999). An interesting construction is the use of hidden Markov models on trees coupled with wavelets in order to induce dependence among the wavelet coefficients (Crouse et al., 1998; Nowak, 1999).

There are other approaches to the analysis of images using multiscale label representations. For example, Gidas (1989) and Pérez and Heitz (1996) define the label field model at the finest level and, given some coarsening operation, obtain approximations for the model at the coarser levels. Another approach based on multigrid methods (for an introduction to multigrid methods, see Briggs et al., 2000) is to initiate the image analysis at the coarsest level, use that analysis as a starting point for the analysis at the second-coarsest level, and proceed analogously until the finest level of resolution. This multigrid-based approach requires the definition of how to coarsen the image and how to interpolate the image analysis results to the next finer level (Bouman and Liu, 1991; Krishnamachari and Chellappa, 1997). These constructions do not lead to coherent joint probability models for the latent label process at the different levels of resolution. We prefer the HMMT framework that provides a coherent joint probability model at the different levels of resolution and, thus, inherits from the laws of probability a natural flow of information between the different resolution levels.

An approach that is related to HMMTs has been proposed by Kolaczyk et al. (2005) for image segmentation. In their approach, conditional on a label field the observations are independent with label-dependent mixture distributions. The label field is obtained through adaptive pruning of recursive dyadic partitions and is estimated by penalized maximum likelihood. The penalization term can be interpreted as the logarithm of a prior that induces parsimonious estimated label fields.

This chapter is organized as follows. Section 8.1 discusses hidden Markov models in one dimension. Section 8.2 presents hidden Markov models on trees. The recursive algorithms for fine to coarse filtering and coarse to fine smoothing are presented in Section 8.3. Estimation when the hyperparameters are unknown is discussed in Section 8.4. We illustrate the use of HMMTs with an application to image classification in Section 8.5.

8.1 HMMs in 1-D

A hidden Markov model in one dimension has two parts: a latent label process that assumes discrete values in a set $B = \{1, \ldots, b\}$ and evolves through time according to a Markov chain; and observations that are (in most models) conditionally independent given the latent label process. This can be seen as a mixture model with a finite state Markov chain as the mixing distribution (e.g., see Scott, 2002). Hidden Markov models in one dimension have been applied in a wide range of problems. For example, Hamilton (1989), Albert and Chib (1993), Chopin and Pelgrin (2004), and Pelletier (2006) have used HMMs to model regime changes in time series useful for detecting economic cycles; in that case, the latent label process may assume distinct discrete values that correspond to different economic regimes. Husmeier and McGuire (2003) have developed an HMM-based model for evolution of species with molecular sequences arising from mixtures of topologies adequate to the detection of recombination in 4-taxa DNA sequence alignments. Other fields where HMMs have been applied include speech recognition (Juang and Rabiner, 1991; Sirigos et al., 2002), network security (Scott, 1999, 2004), alignment of molecular sequences (Liu et al., 1999; Neuwald and Liu, 2004), and longitudinal comparisons of treatments (Scott et al., 2005).

Define $\mathbf{y}_{1:t} = (y_1, \ldots, y_t)'$ and $\mathbf{x}_{1:t} = (x_1, \ldots, x_t)'$. The HMM is defined by the transition kernel for the latent label process from time $t-1$ to time t, by the initial distribution at time 0, and by the distribution of the observation at time t conditional on the latent label process

$$p(x_t = a | \mathbf{x}_{1:(t-1)}, \boldsymbol{\theta}) = p(x_t = a | x_{t-1}, \boldsymbol{\theta}), \tag{8.1}$$

$$p(x_1 = a), \quad a \in B, \tag{8.2}$$

$$p(y_t | \mathbf{x}_{1:t}, \mathbf{y}_{1:(t-1)}, \boldsymbol{\theta}) = p(y_t | x_t, \boldsymbol{\theta}) = f_{x_t}(y_t | \boldsymbol{\theta}), \tag{8.3}$$

where $a \in B$, B is the set of labels, $t = 1, \ldots, T$, and $f_{x_t}(y_t | \boldsymbol{\theta})$ may be a density function or a probability function.

Bayesian analysis of HMMs in 1-D is typically performed within an MCMC framework in two steps. The first step simulates the latent label vector $(x_1, \ldots, x_T)$ from its joint full conditional distribution using a forward filter backward sampler. The second step, which is problem-specific, simulates the parameter vector $\boldsymbol{\theta}$ from its full conditional distribution. Let us now describe the recursive equations for the first step. For the sake of simplicity, we omit the dependence on $\boldsymbol{\theta}$.

The analysis of HMMs starts with the computation of the posterior distribution of the latent variable at time 1 given the observation at time 1:

$$p(x_1 | y_1) = \frac{p(y_1 | x_1) p(x_1)}{\sum_{a=1}^{b} p(y_1 | x_1 = a) p(x_1 = a)}.$$

Then, a forward filter computes

$$p(x_t | \mathbf{y}_{1:(t-1)}) = \sum_{a=1}^{b} p(x_t | x_{t-1} = a) p(x_{t-1} = a | \mathbf{y}_{1:(t-1)}),$$

$$p(x_t | \mathbf{y}_{1:t}) = \frac{f_{x_t}(y_t) p(x_t | \mathbf{y}_{1:(t-1)})}{\sum_{a=1}^{b} f_a(y_t) p(x_t = a | \mathbf{y}_{1:(t-1)})}.$$

The forward filter stops at time T and provides $p(x_T | \mathbf{y}_{1:t})$, the posterior distribution of the latent label process at time T given all the observations. Then, the following backward smoother computes the posterior distributions of the latent label process at each time conditional on all the observations:

$$p(x_t | \mathbf{y}_{1:T}) = \sum_{a=1}^{b} p(x_t | \mathbf{y}_{1:t}, x_{t+1} = a) p(x_{t+1} = a | \mathbf{y}_{1:T}),$$

where

$$p(x_t | \mathbf{y}_{1:t}, x_{t+1} = a) = \frac{p(x_{t+1} = a | x_t) p(x_t | \mathbf{y}_{1:t})}{p(x_{t+1} = a | \mathbf{y}_{1:t})}.$$

The equations above for the forward filter backward smoother may be numerically unstable. Refer to Scott (2002) for modifications that make the recursions more stable.

8.2 HMMs on Trees

The construction of a hidden Markov model on a tree is analogous to the construction of Gaussian multiscale models on trees described in Chapter 7.

More specifically, a hidden Markov model on a tree assumes that the values of the latent label process at nodes of a given level are conditionally independent given the latent label process at the immediate coarser level. The notation here is analogous to that of Chapter 7; that is, $\mathbf{x}_l$ and $\mathbf{y}_l$ denote the vector of labels and the vector of observations at resolution level l.

For the sake of simplicity of exposition, we assume here that the observations at each node and each level are conditionally independent given the latent process. Thus, the joint density function of $(\mathbf{y}_0, \ldots, \mathbf{y}_{L-1}, \mathbf{x}_0, \ldots, \mathbf{x}_{L-1})$ can be written as

$$p(\mathbf{y}_0, \ldots, \mathbf{y}_{L-1}, \mathbf{x}_0, \ldots, \mathbf{x}_{L-1}|\boldsymbol{\theta}) = p(\mathbf{x}_0|\boldsymbol{\theta}) \left[\prod_{l=1}^{L-1} \prod_{j=1}^{n_l} p(x_{lj}|x_{A_{lj}}, \boldsymbol{\theta}) \right]$$

$$\times \left[\prod_{l=0}^{L-1} \prod_{j=1}^{n_l} p(y_{lj}|x_{lj}, \boldsymbol{\theta}) \right], \qquad (8.4)$$

where $p(x_{lj}|x_{A_{lj}}, \boldsymbol{\theta})$ is the discrete probability function of x_{lj} given its ancestor, and $f(y_{lj}|x_{lj}, \boldsymbol{\theta})$ is the density function, either discrete or continuous, of y_{lj} given the corresponding latent variable x_{lj}. Moreover, $\boldsymbol{\theta}$ is the parameter vector associated with the multiscale hidden Markov model of interest and is typically of low dimension. For example, $\boldsymbol{\theta}$ may include transition probabilities. Bayesian estimation of the label process and the parameter vector $\boldsymbol{\theta}$ is easily performed regardless of whether the observations y_{lj} are discrete or continuous.

8.3 Estimation When θ Is Known

When $\boldsymbol{\theta}$ is known, a Viterbi-like algorithm (for information on the Viterbi algorithm, see Viterbi, 1967; Forney, 1973; Scott, 2002) can be used for inference on the multiscale latent label process. This algorithm was independently introduced by Dawid (1992) for the analysis of probabilistic expert systems and by Laferté et al. (1995) for discrete image modeling.

8.3.1 Fine to Coarse Filter

The multiscale construction of the discrete latent process defines a marginal prior for each x_{lj}, $p(x_{lj})$, that can be recursively computed as

$$p(x_{lj}) = \sum_{a=1}^{b} p(x_{lj}|x_{A_{lj}} = a)p(x_{A_{lj}} = a). \qquad (8.5)$$

Moreover, the coarse to fine construction of the latent process can be reversed using Bayes' Theorem. More specifically, if $(l', j') \in D_{lj}$ then

$$p(x_{lj}|x_{l'j'}) = \frac{p(x_{l'j'}|x_{lj})p(x_{lj})}{p(x_{l'j'})}. \tag{8.6}$$

The following theorem provides the fine to coarse filter.

Theorem 8.1. *In the hidden Markov multiscale model defined by Equation (8.4), filtered distributions are given as follows:*

(a) Posterior at the finest resolution level:

$$p(x_{L-1,j}|Y_{L-1,j}) = \frac{p(y_{L-1,j}|x_{L-1,j})p(x_{L-1,j})}{\sum_{a=1}^{b} p(y_{L-1,j}|x_{L-1,j} = a)p(x_{L-1,j} = a)}.$$

(b) Posterior for $(l',j') \in D_{lj}$: Discrete probability distribution

$$p(x_{l'j'}|Y_{l'j'}).$$

(c) Prior at (l,j):

$$p(x_{lj}|Y_{lj}^{+}) \propto p(x_{lj})p(Y_{lj}^{+}|x_{lj})$$
$$= p(x_{lj}) \prod_{(l',j')\in D_{lj}} p(Y_{l'j'}|x_{lj})$$
$$\propto p(x_{lj}) \prod_{(l',j')\in D_{lj}} \frac{p(x_{lj}|Y_{l'j'})}{p(x_{lj})},$$

where $p(x_{lj}|Y_{l'j'})$ can be computed as

$$p(x_{lj}|Y_{l'j'}) = \sum_{a=1}^{b} p(x_{lj}|x_{l'j'} = a)p(x_{l'j'} = a|Y_{l'j'}).$$

(d) Posterior at (l,j):
If $\gamma_{lj} = 0$ then $p(x_{lj}|Y_{lj}) = p(x_{lj}|Y_{lj}^{+})$.
If $\gamma_{lj} = 1$ then $p(x_{lj}|Y_{lj}) \propto p(y_{lj}|x_{lj})p(x_{lj}|Y_{lj}^{+})$.

Proof. The proof is by induction and uses discrete probability calculus and Bayes' Theorem. As the proof is straightforward, we leave it to the reader. □

8.3.2 Coarse to Fine Smoother

After the fine to coarse filtering sweep, we have at the root of the tree the distribution $\mathbf{x}_0|Y_0$; that is, the distribution of the latent label process at the root node given all the information. In order to obtain the smoothed distributions $x_{lj}|Y_0, l = 1, \ldots, L-1, j = 1, \ldots, n_l$, it will be necessary to send the information Y_0 down the tree. Define $Y_{lj}^{-} = Y_0 \setminus Y_{lj}$. The following theorem provides the coarse to fine smoother.

Theorem 8.2. *In the multiscale hidden Markov model defined by Equation (8.4), smoothed distributions are given as follows:*

(a) Posterior for node A_{lj}: Discrete probability distribution $p(x_{A_{lj}}|Y_0)$, $x_{A_{lj}} = 1, \ldots, b$.

(b)

$$p(x_{lj}|x_{A_{lj}}, Y_{lj}) \propto \frac{p(x_{lj}|Y_{lj})p(x_{lj}|x_{A_{lj}})}{p(x_{lj})}.$$

(c) Posterior for node (l, j):

$$p(x_{lj}|Y_0) = \sum_{a=1}^{b} p(x_{lj}|x_{A_{lj}} = a, Y_{lj})p(x_{A_{lj}} = a|Y_0).$$

Proof. Again, the proof is by induction and uses discrete multivariate probability calculus and Bayes' Theorem. As the proof is straightforward, we leave it to the reader. □

8.4 Estimation When θ Is Unknown

If the parameter vector θ is unknown, the Viterbi-like algorithm of Section 8.3 cannot be used directly. Nevertheless, the fine to coarse filtering sweep and the coarse to fine smoothing sweep may be integrated within algorithms for the joint Bayesian or maximum likelihood estimation of the parameter vector θ and the latent process x. As discussed in Section 7.4 for the Gaussian case, maximum likelihood estimation will typically be performed by an EM algorithm, while Bayesian estimation will use a Markov chain Monte Carlo algorithm or an empirical Bayes approach.

Moreover, as θ includes the hyperparameters of the conditional distribution of an observation given the corresponding label, when θ is unknown there is ambiguity in the definition of the model because a switch of two labels will yield the same likelihood. The typical solution for this problem is to impose parameter constraints, such as ordering of the means or variances corresponding to different labels (e.g., see Scott, 2002).

8.5 Application: Image Classification

In this section, we illustrate the use of multiscale hidden Markov models with an application to the classification of a synthetic image.

We consider here a field of size 128×128. Each pixel of this field can be classified as belonging to one of three groups, A, B, or C, as depicted in Figure 8.1(a). Associated with each pixel is a Gaussian observation with parameters that depend on the pixel classification:

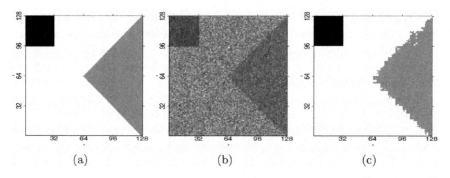

(a)   (b)   (c)

Fig. 8.1. (a) True label field: white (group 1), grey (group 2), black (group 3). (b) Observed noisy field (light colors correspond to higher values, darker colors correspond to lower values). (c) Maximum a posteriori label field: white (group 1), grey (group 2), black (group 3).

- Group 1: $N(\mu_1, \sigma_1^2)$.
- Group 2: $N(\mu_2, \sigma_2^2)$.
- Group 3: $N(\mu_3, \sigma_3^2)$.

Conditional on the label field (that is, the field that contains the classifications of the pixels), the observations are independent. Here we used $\mu_1 = 9$, $\mu_2 = 7$, $\mu_3 = 5$, $\sigma_1^2 = 4$, $\sigma_2^2 = 2$, $\sigma_3^2 = 1$. In the computations below, it is assumed that these parameters are known. The simulated observed field is depicted in Figure 8.1(b).

The objective here is to use the observed field to classify each pixel of the image as belonging to group 1, 2, or 3. In order to do that, we assume a hidden Markov model on a quadtree with $\log_2 128 = 7$ resolution levels. In this particular example, the observations are made only at the finest resolution level (128×128).

It is assumed that there is no prior information on the label at the root level; that is, $p(x_0 = a) = 1/3, a = 1, 2, 3$. Moreover, it is assumed that the transition probability from parent node to children nodes is given by

$$p(x_{lj} = b | x_{A_{lj}} = a) = \begin{cases} 1 - \theta^l, & \text{if } b = a, \\ 0.5\theta^l, & \text{otherwise.} \end{cases}$$

Thus, as we go from coarser to finer levels, the probability that a descendant node assumes the same label as its ancestral node increases. That is a good idea because it induces more similarity between nodes at the finer levels while allowing larger variability at the coarser levels.

The parameter $\theta \in (0, 1)$ controls the smoothness of the process. Note that the relationship between labels of a given level is induced by their relationship with the most recent common ancestor. Thus, if θ is small, then the labels at the descendant node will tend to be the same as the label at the ancestral node, and that will induce more smoothness. If θ is large, then the dependence between the nodes at the finest resolution level will be weaker. Here we use $\theta = 0.7$, which in fact induces negative correlations between the label nodes at level 1 and their ancestor at level 0 but nevertheless induces smoothness at the finest level of resolution.

Figure 8.1(c) presents the maximum a posteriori (MAP) estimate of the label field at the finest resolution. The multiscale hidden Markov model framework was able to classify without error all the pixels corresponding to group 3. While the framework had some difficulty with the classification of the pixels close to the boundary between groups 1 and 2, most of the pixels of those groups were correctly classified.

9

Mass-Balanced Multiscale Models on Trees

In many multiscale problems, the relationship between the different levels of the multiscale process is deterministic. For example, the total volume of petroleum in a given region at a coarser level is the sum of the volumes of petroleum in the corresponding finer regions. This type of property is known as balance of mass. In this chapter, we revise the tree modeling approach of Kolaczyk and Huang (2001) for the analysis of multiscale processes subject to mass balance. Their approach is strongly based on the approach of Allan S. Willsky and his coauthors, presented in Chapter 7, but generalized to cases where the number of descendants is not constant and to include the cases of Gaussian and Poisson observations. The case of Poisson observations is a generalization of the work of Kolaczyk (1999) for multiscale Poisson processes on dyadic trees. As in Willsky's tree model, the multiscale model of Kolaczyk and Huang induces rich covariance structures whose analytic expressions were derived by Louie and Kolaczyk (2004). Building on the work of Kolaczyk (1999), Bouman et al. (2005) developed a multiscale hazard model for survival analysis. Other alternative multiscale models for the analysis of mass-balanced processes have been developed by Huang et al. (2002) and Daoudi et al. (1999) for Gaussian processes. Here we present the developments of Kolaczyk and Huang (2001) because they include analysis not only for Gaussian but also for Poisson processes.

The procedure of Kolaczyk and Huang (2001) may be used as a multiscale smoothing of the observed values in order to obtain a smoothed estimated latent mean process. There are other alternatives in the literature, and the most currently used assume a priori that the latent mean process follows a Gaussian random field (e.g., Besag et al., 1995). This assumption is not appropriate when some areas have mean levels much higher than their surrounding areas. In those cases, the analysis with Gaussian random fields will oversmooth the latent mean process. In contrast, the multiscale analysis of Kolaczyk and Huang (2001) will tend to indicate that at those areas the latent mean level is in fact much higher than in the surrounding areas. This idea has been further developed by Louie and Kolaczyk (2006a,b) in the context of disease mapping in spatial epidemiology and is very effective for the detection of isolated disease clusters.

9.1 Introduction

Assume that interest lies in some underlying continuous process $\{\mu(s) : s \in S\}$ on some domain $S \subset \mathbb{R}^k$, where k is typically less than or equal to 3. Moreover, assume that because of measurement restrictions, data are available only up to a given scale of resolution $L - 1$ on a partition of the domain S, denoted by $\{B_{L-1,1}, \ldots, B_{L-1,n_{L-1}}\}$, with $B_{L-1,j} \in S$, $j = 1, \ldots, n_{L-1}$, $B_{L-1,i} \cap B_{L-1,j} = \emptyset$, $i \neq j$, and $\cup_{j=1}^n B_{L-1,j} = S$.

Kolaczyk and Huang (2001) assume that for each element $B_{L-1,j}$ there is a measurement $y_{L-1,j}$ such that $E(y_{L-1,j}) = \mu_{L-1,j} = \int_{B_{L-1,j}} \mu(s)ds$, $j = 1, \ldots, n_{L-1}$. Moreover, they assume that $y_{L-1,1}, \ldots, y_{L-1,n_{L-1}}$ are conditionally independent given $\mu_{L-1,1}, \ldots, \mu_{L-1,n_{L-1}}$. This is a fairly reasonable assumption equivalent to assuming independent measurement errors. Note that in general the spatial process $\mu(s)$ will be correlated and this spatial dependence will pass over to the measurements $y_{L-1,1}, \ldots, y_{L-1,n_{L-1}}$. Thus, the marginal distribution of the measurements will exhibit spatial dependence.

Kolaczyk and Huang (2001) are interested not only in the process $\mu(s)$ at the resolution scale $L - 1$ but also in the process at aggregated coarser scales. At the lth scale of resolution, the domain S is partitioned in n_l subregions $B_{l,1}, \ldots, B_{l,n_l}$, $l = 0, \ldots, L - 2$. It is assumed that the partition at level l is a refinement of the partition at level $l + 1$; that is, $B_{lj} = \cup_{(l',j') \in D_{lj}} B_{l'j'}$, $l' = l + 1$, where D_{lj} is the set of descendants of subregion j at level l, and $D_{lj} \cap D_{li} = \emptyset, i \neq j$. Note that the number of descendants does not need to be constant; denote the number of descendants of subregion (l, j) by m_{lj}. This construction is very similar to the construction of multiscale models on trees of Chapter 7, except for the property of balance of mass and the fact that the tree nodes correspond here to subregions.

The mass-balance assumption corresponds to the recursive definition at the lth level of aggregated measurements

$$y_{lj} = \sum_{(l',j') \in D_{lj}} y_{l'j'} = \sum_{(l',j') \in D_{lj}} y_{l+1,j'}$$

and aggregated mean process

$$\mu_{lj} = \sum_{(l',j') \in D_{lj}} \mu_{l+1,j'}.$$

Let G be a set of subregions and denote by $\mathbf{y}_G$ and $\boldsymbol{\mu}_G$ the corresponding vectors of measurements and expected values, respectively. Thus $\mathbf{y}_{D_{lj}}$ denotes the vector of descendants of y_{lj}. Note that as y_{lj} is a deterministic function of its descendants, the distribution of $\mathbf{y}_{D_{lj}}$ conditional on y_{lj} is degenerate. In order to resolve this degeneracy, define D_{lj}^* as the set of descendants of y_{lj} with one deleted descendant; which descendant is deleted is arbitrary. Then, the likelihood function admits the following multiscale factorization (Kolaczyk and Huang, 2001):

$$\prod_{j=1}^{n_{L-1}} p(y_{L-1,j}|\mu) = \left[\prod_{j=1}^{n_0} p(y_{0,j}|\mu) \right] \times \prod_{l=0}^{L-2} \prod_{j=1}^{n_l} p(\mathbf{y}_{D_{lj}^*}|y_{lj}, \mu). \qquad (9.1)$$

This factorization holds fairly generally and, as discussed in the following sections, can be specialized in a simple way for the cases of Gaussian and Poisson observations. In the Gaussian case, the conditional distribution of the descendants given the ancestral node is degenerate Gaussian and we use the notation D_{lj}^* in Section 9.2. In the Poisson case, the distribution of the descendants given the ancestral node is multinomial. Even though the multinomial is a degenerate discrete distribution, there is no ambiguity in its definition and thus we use the notation D_{lj} in Section 9.3.

9.2 Gaussian Case

9.2.1 Likelihood

Denote the expected value and variance of y_{lj} by μ_{lj} and σ_{lj}^2, respectively. In addition, define $\mathbf{y}_l = (y_{l,1}, \ldots, y_{l,n_l})'$, $\boldsymbol{\mu}_l = (\mu_{l,1}, \ldots, \mu_{l,n_l})'$, $\boldsymbol{\sigma}_l^2 = (\sigma_{l,1}^2, \ldots, \sigma_{l,n_l}^2)'$, and $\boldsymbol{\Sigma}_l = \mathrm{diag}(\boldsymbol{\sigma}_l^2)$. Under the assumption of conditional independence of $\mathbf{y}_l$ given $\boldsymbol{\mu}_{L-1}$ and $\boldsymbol{\sigma}_{L-1}^2$, the variance at (l, j) can be recursively computed as

$$\sigma_{lj}^2 = \sum_{(l',j') \in D_{lj}} \sigma_{l+1,j'}^2.$$

Let us assume that the joint distribution of the observations at the finest level $L-1$ conditional on the mean process μ is multivariate Gaussian. Thus, $\mathbf{y}_l|\boldsymbol{\mu}_l, \boldsymbol{\Sigma}_l \sim N(\boldsymbol{\mu}_l, \boldsymbol{\Sigma}_l)$. Moreover, the joint distribution of y_{lj} and $\mathbf{y}_{D_{lj}^*}$ is

$$\begin{pmatrix} y_{lj} \\ \mathbf{y}_{D_{lj}^*} \end{pmatrix} \Bigg| \boldsymbol{\mu}_{L-1}, \boldsymbol{\sigma}_{L-1}^2 \sim N \left[\begin{pmatrix} \mu_{lj} \\ \boldsymbol{\mu}_{D_{lj}^*} \end{pmatrix}, \begin{pmatrix} \sigma_{lj}^2 & \left(\boldsymbol{\sigma}_{D_{lj}^*}^2\right)' \\ \boldsymbol{\sigma}_{D_{lj}^*}^2 & \boldsymbol{\Sigma}_{D_{lj}^*} \end{pmatrix} \right],$$

where $\boldsymbol{\mu}_{D_{lj}^*}$ and $\boldsymbol{\Sigma}_{D_{lj}^*}$ are the mean vector and covariance matrix of $\mathbf{y}_{D_{lj}^*}$, respectively. In addition, $\boldsymbol{\sigma}_{D_{lj}^*}^2$ is the cross covariance matrix between $\mathbf{y}_{D_{lj}^*}$ and y_{lj}; in this case, the cross covariance matrix is equal to the vector of variances of $\mathbf{y}_{D_{lj}^*}$. Thus, from standard results for multivariate normal distributions, the conditional distribution of $\mathbf{y}_{D_{lj}^*}$ given y_{lj} is

$$\mathbf{y}_{D^*_{lj}} \mid y_{lj}, \boldsymbol{\mu}_{L-1}, \sigma^2_{L-1} \sim N(\mathbf{v}_{lj} y_{lj} + \boldsymbol{\omega}_{lj}, \boldsymbol{\Omega}_{lj}),$$

with

$$\mathbf{v}_{lj} = \boldsymbol{\sigma}^2_{D^*_{lj}} / \sigma^2_{lj},$$

$$\boldsymbol{\omega}_{lj} = \boldsymbol{\mu}_{D^*_{lj}} - \mathbf{v}_{lj} \mu_{lj},$$

and

$$\boldsymbol{\Omega}_{lj} = \boldsymbol{\Sigma}_{D^*_{lj}} - \sigma^{-2}_{lj} \boldsymbol{\sigma}^2_{D^*_{lj}} \left(\boldsymbol{\sigma}^2_{D^*_{lj}} \right)'.$$

Therefore, all the distributions in Equation (9.1) are also Gaussian, and the factorization may be written as

$$\prod_{j=1}^{n_{L-1}} p(y_{L-1,j} \mid \mu_{L-1,j}, \sigma^2_{L-1,j}) = p(\mathbf{y}_0 \mid \boldsymbol{\mu}_0, \boldsymbol{\Sigma}_0) \prod_{l=0}^{L-2} \prod_{j=1}^{n_l} p(\mathbf{y}_{D^*_{lj}} \mid y_{lj}, \boldsymbol{\omega}_{lj}, \boldsymbol{\Omega}_{lj}).$$
$$(9.2)$$

9.2.2 Prior Distributions

The factorization (9.2) reparameterizes the Gaussian model initially parameterized by $\boldsymbol{\mu}_{L-1}$, the latent process at the finest level, in terms of $(\boldsymbol{\mu}_0, \boldsymbol{\omega}_0, \dots, \boldsymbol{\omega}_{L-2})$, where $\boldsymbol{\omega}_l = (\boldsymbol{\omega}'_{l1}, \dots, \boldsymbol{\omega}'_{l,n_l})'$. This is a multiscale decomposition of the mean process analogous to wavelet decompositions, where $\boldsymbol{\omega}_{lj}$ has a role similar to wavelet coefficients. More specifically, $\boldsymbol{\omega}_{lj}$ explains how the value at node (l, j) is expected to split among its descendants. As is common practice in wavelet analysis, Kolaczyk and Huang (2001) assign independent priors to each element of $(\boldsymbol{\mu}_0, \boldsymbol{\omega}_0, \dots, \boldsymbol{\omega}_{L-2})$.

A natural choice for the mean level at the coarsest resolution is to assume a conjugate prior $\boldsymbol{\mu}_0 \sim N(\mathbf{m}_{\boldsymbol{\mu}_0}, \boldsymbol{\Phi}_0)$, with $\boldsymbol{\Phi}_0$ implying a fairly diffuse distribution. For the $\boldsymbol{\omega}_{lj}$'s there are some possible specifications that imply different levels of homogeneity for the latent process $\mu(s)$.

In one prior specification, they assume conjugate priors

$$\boldsymbol{\omega}_{lj} \mid \boldsymbol{\Phi}_{lj} \sim N(\mathbf{0}, \boldsymbol{\Phi}_{lj}), \tag{9.3}$$

where $\boldsymbol{\Phi}_{lj}, l = 0, \dots, L-1, j = 1, \dots, n_l$ are known covariance matrices. The use of this conjugate prior allows fast computation of summaries of the posterior distribution such as posterior means and variances.

They also consider a prior specification based on the mixture of distributions

$$\boldsymbol{\omega}_{lj} \mid \eta_{lj}, \boldsymbol{\Phi}^{(0)}_{lj}, \boldsymbol{\Phi}^{(1)}_{lj} \sim (1 - \eta_{lj}) N(\mathbf{0}, \boldsymbol{\Phi}^{(0)}_{lj}) + \eta_{lj} N(\mathbf{0}, \boldsymbol{\Phi}^{(1)}_{lj}), \tag{9.4}$$

where $\eta_{lj} \sim \text{Bernoulli}(p_{lj})$, and $\boldsymbol{\Phi}^{(0)}_{lj}$ and $\boldsymbol{\Phi}^{(1)}_{lj}$ are known covariance matrices such that $\boldsymbol{\Phi}^{(0)}_{lj}$ implies a distribution very concentrated around the mean $\mathbf{0}$, while $\boldsymbol{\Phi}^{(1)}_{lj}$ implies a very diffuse distribution. This specification allows fairly inhomogeneous latent processes $\mu(s)$.

9.2.3 Estimation

The estimation of the latent process at the finest level $\boldsymbol{\mu}_{L-1}$ is performed assuming a square error loss. As a consequence, the Bayesian estimator is the posterior mean of $\boldsymbol{\mu}_{L-1}$.

A very interesting feature of Kolaczyk and Huang's construction is that the posterior distribution of the mean level process at the finest level $\boldsymbol{\mu}_{L-1}$ factorizes in terms of the new parameterization:

$$p(\boldsymbol{\mu}_{L-1}|y_{L-1}) = p(\boldsymbol{\mu}_0|\mathbf{y}_0) \prod_{l=0}^{L-2} \prod_{j=1}^{n_l} p(\boldsymbol{\omega}_{lj}|\mathbf{y}_{D_{lj}^*}, y_{lj}). \tag{9.5}$$

Because of the factorization above, the posterior mean of $\boldsymbol{\mu}_{L-1}$ can be recursively computed using the formula

$$\hat{\boldsymbol{\mu}}_{D_{lj}^*} = \hat{\boldsymbol{\omega}}_{lj} + \mathbf{v}_{lj}\hat{\mu}_{lj},$$

where $\hat{\boldsymbol{\omega}}_{lj} = E[\boldsymbol{\omega}_{lj}|\mathbf{y}_{D_{lj}^*}, y_{lj}]$.

The recursion starts with the computation of the posterior mean of the latent process at the coarsest level:

$$\hat{\boldsymbol{\mu}}_0 = (\boldsymbol{\Sigma}_0^{-1} + \boldsymbol{\Phi}_0^{-1})^{-1}(\boldsymbol{\Sigma}_0^{-1}\mathbf{y}_0 + \boldsymbol{\Phi}_0^{-1}\mathbf{m}_{\boldsymbol{\mu}_0}).$$

Note that when the prior for $\boldsymbol{\mu}_0$ is diffuse (that is, $\boldsymbol{\Phi}_0$ is large), the posterior mean of $\boldsymbol{\mu}_0$ will be close to $\mathbf{y}_0$.

When the conjugate Gaussian prior (9.3) is used, the posterior distribution of $\boldsymbol{\omega}_{lj}$ is Gaussian with mean

$$\hat{\boldsymbol{\omega}}_{lj} = \boldsymbol{\Phi}_{lj}(\boldsymbol{\Phi}_{lj} + \boldsymbol{\Omega}_{lj})^{-1}(\mathbf{y}_{D_{lj}^*} - y_{lj}\mathbf{v}_{lj}). \tag{9.6}$$

When the mixture-of-normals prior (9.4) is used, the posterior distribution of $\boldsymbol{\omega}_{lj}$ is also a mixture of normals. In this case, the posterior mean will be

$$\hat{\boldsymbol{\omega}}_{lj} = q_{lj}\hat{\boldsymbol{\omega}}_{lj}^{(1)} + (1 - q_{lj})\hat{\boldsymbol{\omega}}_{lj}^{(0)},$$

where $\hat{\boldsymbol{\omega}}_{lj}^{(0)}$ and $\hat{\boldsymbol{\omega}}_{lj}^{(1)}$ are computed with Equation (9.6) using $\boldsymbol{\Phi}_{lj}^{(0)}$ and $\boldsymbol{\Phi}_{lj}^{(1)}$, respectively, instead of $\boldsymbol{\Phi}_{lj}$. Moreover,

$$q_{lj} = P(\eta_{lj} = 1|\mathbf{y}_{D_{lj}^*}, y_{lj})$$
$$= \frac{O_{lj}}{1 + O_{lj}},$$

where O_{lj} is the posterior odds ratio that node (l, j) belongs to component 1 against component 0 of the mixture; that is,

$$O_{lj} = \frac{p_{lj}}{1 - p_{lj}} \times$$
$$\frac{|\boldsymbol{\Phi}_{lj}^{(0)} + \boldsymbol{\Omega}_{lj}|^{0.5} \exp\left\{0.5(\mathbf{y}_{D_{lj}^*} - y_{lj}\mathbf{v}_{lj})'(\boldsymbol{\Phi}_{lj}^{(0)} + \boldsymbol{\Omega}_{lj})^{-1}(\mathbf{y}_{D_{lj}^*} - y_{lj}\mathbf{v}_{lj})\right\}}{|\boldsymbol{\Phi}_{lj}^{(1)} + \boldsymbol{\Omega}_{lj}|^{0.5} \exp\left\{0.5(\mathbf{y}_{D_{lj}^*} - y_{lj}\mathbf{v}_{lj})'(\boldsymbol{\Phi}_{lj}^{(1)} + \boldsymbol{\Omega}_{lj})^{-1}(\mathbf{y}_{D_{lj}^*} - y_{lj}\mathbf{v}_{lj})\right\}}.$$

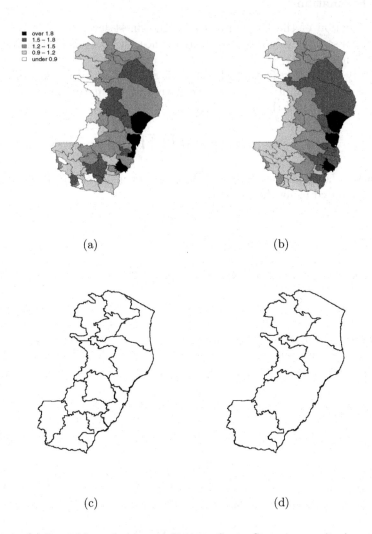

Fig. 9.1. (a) Logarithm of observed Espirito Santo State per capita income per county in 2001. (b) Logarithm of smoothed per capita income. (c) Micro-region map. (d) Macro-region map. The smoothed map highlights the existence of a cluster of higher per capita level in the central-east part of the state that corresponds to the metropolitan area of Vitoria.

9.2.4 Application: Per Capita Income in Espirito Santo State

We illustrate the application of the Gaussian multiscale model to the smoothing of the per capita income per county of Espirito Santo State, Brazil, in 2001 (Figure 9.1). The per capita income is a continuous variable, and for easy visualization Figure 9.1(a) shows in five categories the logarithm of the observed per capita income per county of Espirito Santo State in 2001. Here we consider three levels of resolution: county (fine), micro-region (intermediate), and macro-region (coarse). Here we do not have mass balance as defined in Section 9.1, but we are interested in using the model for smoothing. In this application, the observations are made at the fine level of resolution; that is, at the county level. There is high disparity among counties of Espirito Santo State with respect to per capita income, with highest and lowest per capita income differing by a factor of 3. Figures 9.1(c) and 9.1(d) show the political boundaries of the micro- and macro-regions, respectively.

We applied the Gaussian multiscale model estimating the variances by empirical Bayes. Figure 9.1(b) shows the smoothed per capita income per county. The smoothing is highly related to the definitions of the micro- and macro-regions. Thus, this type of procedure is adequate when the quantity of interest is related to the political boundaries at the several resolution levels. In the case of Espirito Santo State, counties in the same micro- and macro-regions have similar socio-economic characteristics. From Figure 9.1(b), it is interesting to note that counties in the east of Espirito Santo State have higher smoothed per capita incomes. Another important feature is the existence of a cluster of higher per capita levels in the central-east part of the state that corresponds to the metropolitan area of the state capital, Vitoria.

9.3 Poisson Case

9.3.1 Likelihood

When the distribution of the observations at the finest level $L-1$ conditional on the latent process μ is Poisson, then all the distributions in Equation (9.1) will be Poisson and multinomial. More specifically, in the case of Poisson observations, the factorization (9.1) may be written as

$$\prod_{j=1}^{n_{L-1}} p(y_{L-1,j}|\mu_{L-1,j}) = p(\mathbf{y}_0|\boldsymbol{\mu}_0) \prod_{l=0}^{L-2} \prod_{j=1}^{n_l} p(\mathbf{y}_{D_{lj}}|y_{lj},\boldsymbol{\omega}_{lj}), \qquad (9.7)$$

where $\mathbf{y}_0|\boldsymbol{\mu}_0 \sim \text{Poisson}(\boldsymbol{\mu}_0)$ and $\mathbf{y}_{D_{lj}}|y_{lj},\boldsymbol{\omega}_{lj} \sim \text{Multinomial}(y_{lj},\boldsymbol{\omega}_{lj})$, with $\boldsymbol{\omega}_{lj} = \boldsymbol{\mu}_{D_{lj}}/\mu_{lj}$.

9.3.2 Prior Distributions

As in the Gaussian case, the factorization (9.7) reparameterizes the Poisson model initially parameterized by the latent process at the finest level $\boldsymbol{\mu}_{L-1}$ in terms of $(\boldsymbol{\mu}_0, \boldsymbol{\omega}_0, \ldots, \boldsymbol{\omega}_{L-2})$, where $\boldsymbol{\omega}_l = (\boldsymbol{\omega}'_{l1}, \ldots, \boldsymbol{\omega}'_{l,n_l})'$.

A natural choice for the mean level at the coarsest resolution, $\boldsymbol{\mu}_0$, is to assume independent conjugate gamma priors for each element; that is, $\mu_{0j} \sim$

Gamma(τ_0, τ_1), $j = 1, \ldots, n_0$, with $\tau_0 > 0$ and $\tau_1 > 0$. Typically, τ_0 and τ_1 are chosen close to zero, implying a fairly diffuse distribution.

Again, for the $\boldsymbol{\omega}_{lj}$'s there are some possible specifications that imply different levels of homogeneity for the latent process $\mu(s)$. One possible specification is to assume conjugate priors

$$\boldsymbol{\omega}_{lj} | \gamma_{lj} \sim \text{Dirichlet}(\gamma_{lj} \mathbf{1}_{m_{lj}}), \qquad (9.8)$$

where $\gamma_{lj} > 0$ and $\mathbf{1}_{m_{lj}}$ is a vector of ones with length equal to the number of descendants m_{lj} of node (l, j). The use of this conjugate prior allows fast multiscale analysis.

Another possibility is to assume a priori a mixture of distributions,

$$\boldsymbol{\omega}_{lj} | \eta_{lj}, \gamma_{lj}^{(0)}, \gamma_{lj}^{(1)} \sim (1 - \eta_{lj}) \text{Dirichlet}(\gamma_{lj}^{(0)} \mathbf{1}_{m_{lj}}) + \eta_{lj} \text{Dirichlet}(\gamma_{lj}^{(1)} \mathbf{1}_{m_{lj}}), \quad (9.9)$$

where $\eta_{lj} \sim \text{Bernoulli}(p_{lj})$, $\gamma_{lj}^{(0)} > 0$ is small, and $\gamma_{lj}^{(1)} > 0$ is large. In the limiting case $\gamma_{lj}^{(1)} \to \infty$, the second component of the mixture becomes a degenerate multivariate distribution with each element equal to $1/m_{lj}$, which implies an equal split of the value at node (l, j) between its descendants.

9.3.3 Estimation

As in the Gaussian case, in the Poisson case the posterior distribution of the mean level process at the finest level, $\boldsymbol{\mu}_{L-1}$, factorizes as in Equation (9.5). As a consequence, the posterior mean of $\boldsymbol{\mu}_{L-1}$ can be recursively computed using the formula

$$\hat{\boldsymbol{\mu}}_{D_{lj}} = \hat{\boldsymbol{\omega}}_{lj} \hat{\mu}_{lj},$$

where $\hat{\boldsymbol{\omega}}_{lj} = E[\boldsymbol{\omega}_{lj} | \mathbf{y}_{D_{lj}}, y_{lj}]$.

The recursion starts with the computation of the posterior mean of the latent process at the coarsest level. The posterior distribution of μ_{0j} is Gamma$(y_{0j} + \tau_0, 1 + \tau_1)$. Thus, the posterior mean is

$$\hat{\mu}_{0j} = \frac{y_{0j} + \tau_0}{1 + \tau_1}.$$

Note that when the prior for $\boldsymbol{\mu}_0$ is diffuse (that is, τ_0 and τ_1 are close to zero), the posterior mean of $\boldsymbol{\mu}_0$ will be close to $\mathbf{y}_0$.

When the conjugate Dirichlet prior (9.8) is used, the posterior distribution of $\boldsymbol{\omega}_{lj}$ is Dirichlet$(\gamma_{lj} \mathbf{1}_{m_{lj}} + \mathbf{y}_{D_{lj}})$. In this case, the posterior mean of $\boldsymbol{\omega}_{lj}$ is

$$\hat{\boldsymbol{\omega}}_{lj} = \frac{\gamma_{lj} \mathbf{1}_{m_{lj}} + \mathbf{y}_{D_{lj}}}{\gamma_{lj} m_{lj} + y_{lj}}. \qquad (9.10)$$

When the mixture of Dirichlets prior (9.9) is used, the posterior distribution of $\boldsymbol{\omega}_{lj}$ is also a mixture of Dirichlets. In this case, the posterior mean will be

$$\hat{\boldsymbol{\omega}}_{lj} = q_{lj}\hat{\boldsymbol{\omega}}_{lj}^{(1)} + (1 - q_{lj})\hat{\boldsymbol{\omega}}_{lj}^{(0)},$$

where $\hat{\boldsymbol{\omega}}_{lj}^{(0)}$ and $\hat{\boldsymbol{\omega}}_{lj}^{(1)}$ are computed with Equation (9.10) using $\gamma_{lj}^{(0)}$ and $\gamma_{lj}^{(1)}$, respectively, instead of γ_{lj}.

Moreover, q_{lj} is the posterior probability that node (l, j) belongs to component 1 of the mixture of Dirichlets,

$$q_{lj} = P(\eta_{lj} = 1 | \mathbf{y}_{D_{lj}}, y_{lj})$$
$$= \frac{O_{lj}}{1 + O_{lj}},$$

where O_{lj} is the posterior odds ratio of node (l, j) belonging to component 1 rather than component 0 of the mixture,

$$O_{lj} = \frac{p_{lj}}{1 - p_{lj}}$$
$$\times \frac{p(\mathbf{y}_{D_{lj}} | y_{lj}, \eta_{lj} = 1)}{p(\mathbf{y}_{D_{lj}} | y_{lj}, \eta_{lj} = 0)}, \tag{9.11}$$

where $p(\mathbf{y}_{D_{lj}} | y_{lj}, \eta_{lj} = c)$ is the predictive distribution of the descendants $\mathbf{y}_{D_{lj}}$ conditional on the parent y_{lj} and on node (l, j) belonging to component c of the mixture. Standard probability results show that this predictive distribution can be computed as

$$p(\mathbf{y}_{D_{lj}} | y_{lj}, \eta_{lj} = c) = \frac{\Gamma\left(m_{lj}\gamma_{lj}^{(c)}\right)}{\Gamma\left(\gamma_{lj}^{(c)}\right)^{m_{lj}}} \frac{\prod_{(l',j') \in D_{lj}} \Gamma\left(\gamma_{lj}^{(c)} + y_{l'j'}\right)}{\Gamma\left(m_{lj}\gamma_{lj}^{(c)} + y_{lj}\right)}. \tag{9.12}$$

9.3.4 Application: Smoothing of Leptospirosis Time Series

We illustrate here the use of the multiscale Poisson model with an application to the series of leptospirosis patients admitted to the Couto Maia Hospital from March 15, 1996 to March 15, 2004 in the city of Salvador, Bahia, Brazil. This dataset has been analyzed in more detail by Bustamante et al. (2006). The Couto Maia Hospital receives about 95% of all leptospirosis cases in Salvador. Leptospirosis is a potentially fatal disease that is epidemic in Salvador and has been associated in that city with floods and poor sanitation in slums (Ko et al., 1999; Sarkar et al., 2002). Here we consider the number of cases reported in periods of 11 days, leading to a time series with 256 observations presented in Figure 9.2(a).

We applied the multiscale Poisson model with dyadic partition of Kolaczyk (1999) and $1 + \log_2 256 = 9$ scales. This model uses a tree mixing scheme that implies a stationary multiscale model. We assumed a degenerate mixture prior (9.9) for $\boldsymbol{\omega}_{lj}$ with $\gamma_{lj}^{(0)} = 1$ and $\gamma_{lj}^{(1)} \to \infty, \forall l, j$. Moreover, we used equal

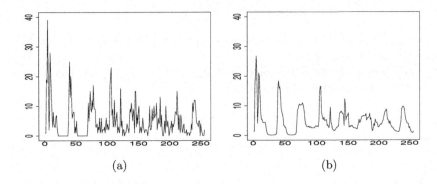

(a)   (b)

Fig. 9.2. (a) Series of leptospirosis from March 15, 1996 to March 15, 2004 aggregated in periods of 11 days. (b) Estimated intensity. The estimated intensity draws attention to the existence of a seasonal pattern and a reduction in the seasonal fluctuations through time.

mixing probabilities at each level; that is, $p_{lj} = p_l$. Additionally, we assumed that the p_l's follow an a priori independent uniform distribution on $[0, 1]$. The mixing probabilities were estimated with marginal posterior modes and the intensity function was estimated as explained in Section 9.3.3.

Figure 9.2(b) presents the estimated intensity function. From the estimated intensity, we can note a seasonal pattern and a reduction in the seasonal fluctuations along the considered time period. While surveillance efforts probably reduced the maximum annual risk over the years, deteriorating urban conditions seem to have led to an increase in the minimum annual risk.

Multiscale Random Fields

This chapter presents multiscale random field models with interconnected resolution levels and smoothness within each level that were introduced by Ferreira et al. (2005). This is important when interest lies in modeling the dynamics of the multiscale process within each scale of resolution as well as across the different scales. This is of particular interest when the process is observed at different scales of resolution or when the multiscale model is used as a prior for an underlying process, and it is desirable to impose smoothness within each level of resolution and a stochastic constraint that the different levels of resolution have similar behavior.

For example, Figure 10.1 represents a multiscale random field with three levels of resolution. At the coarsest and intermediate levels of resolution, each site corresponds to four sites of the immediate finer level. It is desirable to impose some stochastic structure at each level, such as Markov random fields, and connect the levels through some coarsening or aggregation operation, such as averaging or sampling, but in general this leads to incompatible probability distributions. Jeffrey's rule of conditioning is used to revise the implied distributions and ensure that the probability distributions at different levels are strictly compatible.

As with the Gaussian multiscale models on trees presented in Chapter 7, the multiscale random field models presented in this chapter have the ability to accommodate observations at different levels of resolution. An advantage of the multiscale random field models of this chapter over the models presented in Chapter 7 is that they provide smoother estimates of the process of interest. An example where multiscale random field models have been used with success is in the estimation of permeability fields. In that problem, there are two types of information: static data and dynamic data. Static data are available at different resolution levels as a result of geological studies, well tests, and core samples. Dynamic data are obtained from results of tracer experiments performed in an aquifer or from the history of oil production in a petroleum field. The dynamic data can be incorporated in the estimation of the permeability field with the help of fluid flow simulators that can run at different

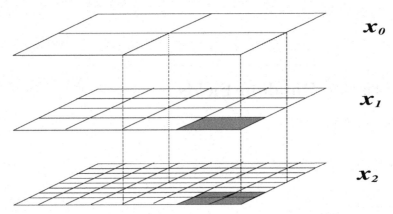

Fig. 10.1. Three levels of a multiscale random field. A particular region at a given resolution level is connected to its neighbors at the same level, to its parent at the immediate coarser level, and to its descendants at the immediate finer level.

resolution levels. Practical implementations use coarser-scale runs that are fast and finer-scale runs that are more accurate. The multiscale random field models of this chapter provide a natural framework for the incorporation of the static data available at different levels as well as for the incorporation of the dynamic data with the use of fluid flow simulators running at different resolution levels. Section 16.2 of Chapter 16 gives a detailed description of the use of multiscale random fields for the estimation of permeability fields.

Let us now introduce some notation. In general, a multiscale random field with L levels of resolution is considered, where $\mathbf{x}_l$ denotes the vectorized lth level of resolution with n_l elements, $l = 0, \ldots, L - 1$; x_{lj} denotes the value of the jth site (or region) of the lth level and (l, j) its corresponding index. It is assumed that site (l, j) has b_{lj} neighbor sites at the same level of resolution, one parent site at the immediate coarser level $l - 1$, and $m_{l+1,j}$ descendant sites at the immediate finer level $l + 1$. Moreover, N_{lj} denotes the set of sites of the lth level that are neighbors of (l, j), and D_{lj} denotes the set of sites of the $(l + 1)$th level that are descendants of (l, j).

The remainder of this chapter is organized as follows. Section 10.1 presents the basic ideas of the construction of multiscale random fields through the construction of a two-level model. This class of models is generalized to an arbitrary number of levels in Section 10.2. Section 10.5 describes the algorithmic implementation of the model by means of a Markov chain Monte Carlo (MCMC) sampling scheme. Section 10.6 illustrates with a simulated dataset the procedures of multiscale random field simulation and parameter estimation.

10.1 Two-Level Model

In order to introduce the main concepts of multiscale random field modeling, this section presents the construction of a two-level multiscale random field model. We refer to the two-levels as the coarse and fine scales.

The definition of the model is somehow subtle — it is important to carefully consider the information on which each equation is conditioned; otherwise the model will be inconsistent. In order to build a consistent model with smoothness at each resolution level and link between the levels, we use here the concept of probability kinematics and the associated Jeffrey's rule of conditioning (for detailed information on probability kinematics and Jeffrey's rule, see Jeffrey, 1988). This rule explains how to revise an old joint probability model for unknowns when new information completely revises the marginal probability distribution for a subset of these unknowns. In our case, we build an initial joint probability model for the coarse and fine levels by assuming that the fine level follows a Markov random field process and by assuming that the coarse level is a linear function of the fine level plus noise. We note that this in general leads to weak spatial correlation at the coarse level and, thus, the coarse-level process would not be very smooth. But let us assume that we believe that the coarse level follows a smooth process, and we incorporate this belief into the model by revising the marginal probability distribution of the coarse level and assuming that the coarse level follows an MRF process. This belief is the new information that completely revises the marginal probability model at the coarse level. But this new information is at odds with the old information that led us to assume an MRF at the fine level and a linear link between levels. Jeffrey's rule is then used to revise the old joint probability distribution at coarse and fine levels. Using Jeffrey's rule, we assume that the conditional distribution of the process at the fine level given the coarse level remains the same. Thus the joint density updated by Jeffrey's rule is the product of the conditional density of the fine level given the coarse level with the revised density of the coarse level. Finally, the updated marginal distribution at the fine level is obtained from the updated joint distribution by marginalization.

To be more concrete, start by assuming that the fine level $\mathbf{x}_1 = (x_{11}, \ldots, x_{1n_1})$ can be modeled by a proper Markov random field (PMRF) model as in Equation (2.1),

$$\mathbf{x}_1 \sim N(\mu \mathbf{1}_{n_1}, \boldsymbol{\Sigma}_1), \tag{10.1}$$

with $\mu \in \mathbb{R}$ a location parameter and $\boldsymbol{\Sigma}_1$ a covariance matrix. As before, we take $\boldsymbol{\Sigma}_1^{-1} = \tau_1(\alpha_1 \mathbf{I}_{n_1} + \mathbf{H}_1)$, with $\tau_1 > 0$ a scale parameter. $\mathbf{H}_1$ is a matrix defining the spatial similarities at the fine level,

$$(\mathbf{H}_1)_{kl} = \begin{cases} h_{1k}, & k = l, \\ -g_{1kl}, & (1, k) \in N_{1l}, \\ 0, & \text{otherwise,} \end{cases} \tag{10.2}$$

with $g_{1kl} > 0$ being a "measure of similarity" between sites $(1, k)$ and $(1, l)$ and $h_{1k} = \sum_{l \in N_{1k}} g_{1kl}$.

As will be made clear, the positive definiteness of Σ_1^{-1} and thus the existence of Σ_1 are fundamental for this particular multiscale model construction. In order to guarantee that Σ_1^{-1} is positive definite, it is assumed that $\alpha_1 > 0$ since that implies Σ_1^{-1} is diagonally dominant and therefore positive definite (Harville, 1997).

Assume a coarsening or aggregation operation through a linear link equation that establishes that the value at each site at the coarse level is equal to a weighted average of the corresponding sites at the fine level plus an error term. More specifically, the link equation is defined as

$$p(\mathbf{x}_0|\mathbf{x}_1) = N(\mathbf{x}_0|\mathbf{A}_1\mathbf{x}_1, \delta_1\mathbf{I}_{n_0}), \tag{10.3}$$

where A_1 is the matrix that performs the coarsening operation and δ_1 is the variance of the error term. Thus, the sum of the elements of each line of $\mathbf{A}_1$ is equal to one. For example, the term $\mathbf{A}_1\mathbf{x}_1$ can represent arithmetic block averages or sampling of the fine level $\mathbf{x}_1$. If arithmetic block averages are considered, then Equation (10.3) can be rewritten as

$$p(\mathbf{x}_0|\mathbf{x}_1) = \prod_{i=1}^{n_0} N\left(x_{0i}\middle| m_1^{-1}\sum_{j \in D_{0i}} x_{1j}, \delta_1\right), \tag{10.4}$$

where m_l is the number of descendants at the fine level of each site at the coarse level and D_{0i} is the set of descendants of the ith coarse-level site. The PMRF model at the fine level $\mathbf{x}_1$ and the link Equation (10.3) imply the particular model $p(\mathbf{x}_0) = N(\mathbf{x}_0|\mu\mathbf{1}_{n_0}, \mathbf{A}_1\Sigma_1\mathbf{A}_1' + \delta_1\mathbf{I}_{n_0})$ for the coarse level. As discussed by Lakshmanan and Derin (1993), in general the Markovianity is lost in this coarsening operation. Thus, the resulting model at the coarse level is more complex than desired and does not lead to efficient algorithms for incorporating information at different resolutions. In order to deal with this problem, Lakshmanan and Derin (1993) suggested approximating the process at the coarse level by a Markov random field. However, in that case the joint probabilistic model of fine and coarse levels becomes inconsistent. Ferreira et al. (2005) suggest another route: part of the information contained in Equations (10.1) and (10.3) is revised in order to impose a simple model at the coarse level.

Suppose that additional information about the coarse level $\mathbf{x}_0$ is received, information that supersedes the prior information on which $p(\mathbf{x}_0)$ is based and directly revises $p(\mathbf{x}_0)$ to an updated model $q(\mathbf{x}_0)$. For example, suppose $q(\mathbf{x}_0)$ is a PMRF,

$$q(\mathbf{x}_0) = N(\mathbf{x}_0|\mu\mathbf{1}_{n_0}, \mathbf{Q}_0), \tag{10.5}$$

where $\mathbf{Q}_0^{-1} = \tau_0[\alpha_0\mathbf{I}_{n_0} + \mathbf{H}_0]$, and

$$\{\mathbf{H}_0\}_{kl} = \left\{ \begin{array}{ll} h_{0k}, & k = l, \\ -g_{0kl}, & (0, k) \in N_{0l}, \\ 0, & \text{otherwise.} \end{array} \right.$$

where the interpretation of the parameters here is analogous to the interpretation of the parameters in Equations (10.1) and (10.2).

The revision of the model for the coarse level from $p(\mathbf{x}_0)$ to $q(\mathbf{x}_0)$ can be viewed as Bayesian updating with an implicit likelihood function proportional to $q(\mathbf{x}_0)/p(\mathbf{x}_0)$. It is assumed that in the revision of beliefs the coarse level $\mathbf{x}_0$ is sufficient for the fine level $\mathbf{x}_1$, where this meaning of sufficiency is as defined by Diaconis and Zabell (1982); i.e., that conditional on the values at the coarse level, the fine level is independent of the additional information that updated the distribution of the coarse level. As a consequence, the conditional distribution of the fine level given the coarse level is not updated by the additional information, and Jeffrey's rule of conditioning can be applied to revise the marginal distribution of $\mathbf{x}_1$ (for details, see Jeffrey, 1988; Diaconis and Zabell, 1982). The following theorem establishes the resulting multiscale model with two-levels of resolution.

Theorem 10.1. *Consider the initial model (10.1) for the fine level* $\mathbf{x}_1$*, the link equation (10.3) and the revised model (10.5) for the coarse level* $\mathbf{x}_0$*. If in the revision of beliefs the coarse level* $\mathbf{x}_0$ *is sufficient for the fine level* $\mathbf{x}_1$*, then*

(i) the joint multiscale model for coarse and fine levels is

$$q(\mathbf{x}_0, \mathbf{x}_1) = N(\mathbf{x}_1 | \mu \mathbf{1}_{n_1} + \mathbf{B}_1(\mathbf{x}_0 - \mu \mathbf{1}_{n_0}), \boldsymbol{\Sigma}_1 - \mathbf{B}_1 \mathbf{W}_1 \mathbf{B}_1') N(\mathbf{x}_0 | \mu \mathbf{1}_{n_0}, \mathbf{Q}_0); \tag{10.6}$$

(ii) the revised marginal model for the fine level is

$$q(\mathbf{x}_1) = N(\mathbf{x}_1 | \mu \mathbf{1}_{n_1}, \boldsymbol{\Sigma}_1 - \mathbf{B}_1(\mathbf{W}_1 - \mathbf{Q}_0)\mathbf{B}_1'), \tag{10.7}$$

where $\mathbf{B}_1 = \boldsymbol{\Sigma}_1 \mathbf{A}_1' \mathbf{W}_1^{-1}$ *and* $\mathbf{W}_1 = \mathbf{A}_1 \boldsymbol{\Sigma}_1 \mathbf{A}_1' + \delta_1 \mathbf{I}_{n_0}$.

Proof.

The assumption that in the revision of beliefs the coarse level $\mathbf{x}_0$ is sufficient for the fine level $\mathbf{x}_1$ means that, conditional on $\mathbf{x}_0$, $\mathbf{x}_1$ is independent of the new information that led to the revision of beliefs about $\mathbf{x}_0$; that is, $q(\mathbf{x}_1 | \mathbf{x}_0) = p(\mathbf{x}_1 | \mathbf{x}_0)$. This last density function can be obtained by Bayes' Theorem:

$$p(\mathbf{x}_1 | \mathbf{x}_0) \propto p(\mathbf{x}_1) p(\mathbf{x}_0 | \mathbf{x}_1)$$
$$= N(\mathbf{x}_1 | \mu \mathbf{1}_{n_1}, \boldsymbol{\Sigma}_1) \, N(\mathbf{x}_0 | \mathbf{A}_1 \mathbf{x}_1, \delta_1 \mathbf{I}_{n_0}).$$

Therefore

$$p(\mathbf{x}_1 | \mathbf{x}_0) = N(\mathbf{x}_1 | \mu \mathbf{1}_{n_1} + \mathbf{B}_1(\mathbf{x}_0 - \mu \mathbf{1}_{n_0}), \boldsymbol{\Sigma}_1 - \mathbf{B}_1 \mathbf{W}_1 \mathbf{B}_1'),$$

where $\mathbf{B}_1 = \boldsymbol{\Sigma}_1 \mathbf{A}_1' \mathbf{W}_1^{-1}$ and $\mathbf{W}_1 = \mathbf{A}_1 \boldsymbol{\Sigma}_1 \mathbf{A}_1' + \delta_1 \mathbf{I}_{n_0}$. Thus, the revised joint multiscale model for coarse and fine levels is

$$q(\mathbf{x}_0, \mathbf{x}_1) \propto q(\mathbf{x}_1 | \mathbf{x}_0) q(\mathbf{x}_0)$$
$$N(\mathbf{x}_1 | \mu \mathbf{1}_{n_1} + \mathbf{B}_1(\mathbf{x}_0 - \mu \mathbf{1}_{n_0}), \boldsymbol{\Sigma}_1 - \mathbf{B}_1 \mathbf{W}_1 \mathbf{B}_1') N(\mathbf{x}_0 | \mu \mathbf{1}_{n_0}, \mathbf{Q}_0).$$

This concludes part (i).

Moreover, the revised marginal model for the fine level can be obtained by the integration of the joint density above with respect to the coarse level, and this is equivalent to the application of Jeffrey's rule of conditioning. Thus

$$q(\mathbf{x}_1) = \int q(\mathbf{x}_0, \mathbf{x}_1) d\mathbf{x}_0 = N(\mathbf{x}_1 | \mu \mathbf{1}_{n_1}, \mathbf{Q}_1),$$

where $\mathbf{Q}_1 = \boldsymbol{\Sigma}_1 - \mathbf{B}_1(\mathbf{W}_1 - \mathbf{Q}_0)\mathbf{B}_1'$. This concludes part (ii). $\square$

10.2 Model with Several Levels

The construction of the multiscale model can be generalized to any arbitrary number of levels through the use of a result by Skyrms (1980), who has shown that the same effect of Jeffrey's rule can be obtained by using a sufficiently richer sample space and by carefully conditioning each equation on different information. In addition, conditioning each equation on different information provides a different interpretation and enlightens interesting aspects of the multiscale model.

Consider a multiscale random field model with L levels of resolution. The lth level is denoted by $\mathbf{x}_l, l = 0, 1, \ldots, L - 1$, with $\mathbf{x}_0$ being the coarsest level, and increases in l meaning progressively finer levels. The movement from coarser to finer levels can be interpreted as an increase in the resolution in a similar interpretation as in the wavelet framework, with the distinction that in this multiscale framework each level can have a meaningful practical interpretation.

The information on which each equation is conditioned is of extreme importance, as the equations would be incompatible if they were conditioned on the same information. Here I_l denotes the initial knowledge about the behavior on the lth level, and G_l is the accumulated knowledge from the coarsest level up to the lth level $\mathbf{x}_l$. More specifically, $G_0 = I_0$ denotes the knowledge about the generator mechanism of the coarsest level $\mathbf{x}_0$ and the accumulated knowledge up to level $\mathbf{x}_0$ as well. In addition, $G_1 = I_0 \bigcap I_1$. In general, the accumulated knowledge from the coarsest level up to the lth level is recursively defined as $G_l = G_{l-1} \bigcap I_l$.

The general multiscale random field model is defined in a hierarchical way. Although it seems that the model is defined from fine to coarse levels, conditional on all the information used to construct the model, the true dependence direction in the model is from coarser to finer levels. That is, following the model definition from fine to coarse levels, the multiscale model is defined as the result of propagating down from coarse to fine levels the implication of revising distributions at the coarse levels. Figure 10.2 presents a partial graphical representation of the model with the true dependence direction from coarser to finer levels.

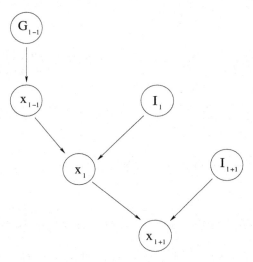

Fig. 10.2. Partial graphical representation of the multiscale random field model.

To be concrete, assume that given I_l, the initial information about level l, the level $\mathbf{x}_l$ follows a proper Markov random field process. Thus

$$\mathbf{x}_l | I_l \sim N(\mu \mathbf{1}_{n_l}, \Sigma_l), l = 0, \ldots, L-1, \tag{10.8}$$

where $\Sigma_l^{-1} = \tau_l(\alpha_l \mathbf{I}_{n_l} + \mathbf{H}_l)$, and the roles and definitions of τ_l, α_l, and $\mathbf{H}_l$ are analogous to the roles and definitions of τ_1, α_1, and $\mathbf{H}_1$ in Section 10.1.

In order to develop estimation and simulation methodologies, it is useful to rewrite Equation (10.8) as (Ferreira and de Oliveira, 2007)

$$p(\mathbf{x}_l | I_l) \propto \tau_l^{0.5n} \prod_{k=1}^{n} (\lambda_{lk} + \alpha_l)^{0.5} \exp \left\{ -0.5\tau_l \left[\mu^2 n_l \alpha_l - 2\mu \alpha_l \sum_{k=1}^{n} x_{lk} \right. \right.$$
$$\left. \left. + \alpha_l \sum_{k=1}^{n} x_{lk}^2 + \mathbf{x}_l' \mathbf{H}_l \mathbf{x}_l \right] \right\}, \tag{10.9}$$

where $\lambda_{l1} \geq \lambda_{l2} \geq \ldots \geq \lambda_{l,n-1} > \lambda_{ln} = 0$ are the eigenvalues of $\mathbf{H}_l$. This expression facilitates the derivation of the full conditional distributions presented in Section 10.5. Moreover, as $\mathbf{H}_l, l = 0, \ldots, L-1$ are assumed known a priori, their eigenvalues need to be computed only once, in the beginning of the estimation procedure. Thus, the use of Equation (10.9) leads to fast computation of the densities of interest in the estimation procedure presented in Section 10.5.2.

In addition, assume that conditional on I_l there exists a known linear transformation mapping the level $\mathbf{x}_l$ to the immediate coarser level $\mathbf{x}_{l-1}$ plus noise. Then, for $l = 1, \ldots, L-1$,

$$p(\mathbf{x}_{l-1}|I_l, \mathbf{x}_l) = N(\mathbf{x}_{l-1}|\mathbf{A}_l\mathbf{x}_l, \mathbf{U}_l), \qquad (10.10)$$

where $\mathbf{U}_l = \delta_l\mathbf{I}_{n_{l-1}}$ and the sum of the elements of each line of $\mathbf{A}_l$ is equal to one. This representation is useful when the relation between coarser and finer levels occurs in a nonregular grid and the number of descendants or the weights are not constant, as in Chapter 9.

In the simplest case, each element of $\mathbf{x}_{l-1}$ is written as an arithmetic average of its m_l descendants at level l plus noise. That is,

$$x_{l-1,s} = m_l^{-1} \sum_{w \in D_{l-1,s}} x_{lw} + u_{l-1,s}, \qquad (10.11)$$

where the $u_{l-1,s}$, $(l = 1, \ldots, L)$, $(s = 1, \ldots, n_l)$, are mutually uncorrelated zero-mean, normally distributed noise terms with $u_{i-l,s} \sim N(u_{l-1,s}|0, \delta_l)$ for some *between-levels* variance δ_l.

Integrating out the finer level $\mathbf{x}_l$, the distribution of the coarser level $\mathbf{x}_{l-1}$ given I_l is obtained as $p(\mathbf{x}_{l-1}|I_l) = N(\mathbf{x}_{l-1}|\mu\mathbf{1}_{n_{l-1}}, \mathbf{A}_l\mathbf{\Sigma}_l\mathbf{A}_l' + \mathbf{U}_l)$. This is straightforward to show since all the involved distributions are Gaussian.

Now, assume that from coarser levels in the hierarchy updated information G_{l-1} is received about the generator mechanism of the coarser level. Following this information, the coarser level $\mathbf{x}_{l-1}$ follows a Gaussian process with mean $\mu\mathbf{1}_{n_{l-1}}$ and covariance matrix $\mathbf{Q}_{l-1}$; that is, $p(\mathbf{x}_{l-1}|G_{l-1}) = N(\mathbf{x}_{l-1}|\mu\mathbf{1}_{n_{l-1}}, \mathbf{Q}_{l-1})$.

As the new information goes down the hierarchy, it is necessary to revise the information on the process at the lth resolution level. Analogously to the construction of the two-level model of Section 10.1, the basic principle is that the accumulated knowledge G_l supersedes the initial knowledge I_l on which the initial distribution for $\mathbf{x}_l$ was based, and is then sent down to finer levels to update the joint distribution throughout the hierarchy.

In order to obtain the joint multiscale model for $(\mathbf{x}_0, \ldots, \mathbf{x}_{L-1})$ given all the knowledge G_{L-1}, the following two hypotheses are assumed:

(1) $p(\mathbf{x}_0, \ldots, \mathbf{x}_{l-1}|G_l) = p(\mathbf{x}_0, \ldots, \mathbf{x}_{l-1}|G_{l-1}, I_l) = p(\mathbf{x}_0, \ldots, \mathbf{x}_{l-1}|G_{l-1})$.
(2) $p(\mathbf{x}_l|G_l, \mathbf{x}_0, \ldots, \mathbf{x}_{l-1}) = p(\mathbf{x}_l|G_{l-1}, I_l, \mathbf{x}_{l-1}) = p(\mathbf{x}_l|I_l, \mathbf{x}_{l-1})$.

Hypothesis (1) states that the coarser levels $\mathbf{x}_0, \ldots, \mathbf{x}_{l-1}$ are independent of the generator mechanism of the finer level $\mathbf{x}_l$ given their own generator mechanism; this hypothesis is equivalent to the subjective update of the distribution of $\mathbf{x}_0, \ldots, \mathbf{x}_{l-1}$.

Hypothesis (2) states that given the generator mechanism of the finer level, the revised information G_{l-1} on the generator mechanism of the coarser level passes to the finer level $\mathbf{x}_l$ only through the realized immediate coarser level $\mathbf{x}_{l-1}$; this hypothesis is equivalent to a Markovian assumption in addition to the sufficiency assumption necessary for the application of Jeffrey's rule.

The following theorem establishes the resulting multiscale model with L levels of resolution.

Theorem 10.2. *The model defined by Equations (10.8) and (10.10) and by Hypotheses (1) and (2) has the following properties:*

(i) the joint multiscale model for $(\mathbf{x}_0, \ldots, \mathbf{x}_{L-1})$ *given all the knowledge* G_{L-1} *is*

$$p(\mathbf{x}_0, \ldots, \mathbf{x}_{L-1}|G_{L-1}) = N(\mathbf{x}_0|\mu\mathbf{1}_{n_0}, \mathbf{Q}_0)$$
$$\prod_{l=1}^{L-1} N(\mathbf{x}_l|\mu\mathbf{1}_{n_l} + \mathbf{B}_l(\mathbf{x}_{l-1} - \mu\mathbf{1}_{n_{l-1}}), \mathbf{\Sigma}_l - \mathbf{B}_l\mathbf{W}_l\mathbf{B}_l');$$

(ii) the marginal model for $\mathbf{x}_l, l = 1, \ldots, L-1$ *given all the knowledge* G_{L-1} *is*

$$p(\mathbf{x}_l|G_{L-1}) = N(\mathbf{x}_l|\mu\mathbf{1}_{n_l}, \mathbf{Q}_l);$$

(iii) and the conditional model of $\mathbf{x}_l$ *given the knowledge from the coarsest up to the* $(L-1)$*th resolution level and the realized process from the coarsest up to the* $(l-1)$*th level is*

$$p(\mathbf{x}_l|G_{L-1}, \mathbf{x}_0, \ldots, \mathbf{x}_{l-1}) = N(\mathbf{x}_l|\mu\mathbf{1}_{n_l} + \mathbf{B}_l(\mathbf{x}_{l-1} - \mu\mathbf{1}_{n_{l-1}}), \mathbf{\Sigma}_l - \mathbf{B}_l\mathbf{W}_l\mathbf{B}_l'),$$

where $\mathbf{B}_l = \mathbf{\Sigma}_l\mathbf{A}_l'\mathbf{W}_l^{-1}$, $\mathbf{W}_l = \mathbf{A}_l\mathbf{\Sigma}_l\mathbf{A}_l' + \delta_l\mathbf{I}_{n_{l-1}}$, *and* $\mathbf{Q}_l = \mathbf{\Sigma}_l - \mathbf{B}_l(\mathbf{W}_l - \mathbf{Q}_{l-1})\mathbf{B}_l'$.

Proof.
Proving part (iii).
Hypothesis (1) implies that $p(\mathbf{x}_0, \ldots, \mathbf{x}_{l-1}|G_{L-1}) = p(\mathbf{x}_0, \ldots, \mathbf{x}_l|G_l)$, and thus

$$p(\mathbf{x}_l|G_{L-1}, \mathbf{x}_0, \ldots, \mathbf{x}_{l-1}) = p(\mathbf{x}_l|G_l, \mathbf{x}_0, \ldots, \mathbf{x}_{l-1})$$
$$= p(\mathbf{x}_l|I_l, \mathbf{x}_{l-1})$$

by Hypothesis (2). Using Bayes' Theorem and Equations (10.8) and (10.10),

$$p(\mathbf{x}_l|I_l, \mathbf{x}_{l-1}) \propto p(\mathbf{x}_l|I_l)p(\mathbf{x}_{l-1}|\mathbf{x}_l, I_l)$$
$$= N(\mathbf{x}_l|\mu\mathbf{1}_{n_l}, \mathbf{\Sigma}_l)N(\mathbf{x}_{l-1}|\mathbf{A}_l\mathbf{x}_l, \mathbf{U}_l).$$

Thus, through Bayesian linear regression,

$$p(\mathbf{x}_l|G_{L-1}, \mathbf{x}_0, \ldots, \mathbf{x}_{l-1}) = N(\mathbf{x}_l|\mu\mathbf{1}_{n_l} + \mathbf{B}_l(\mathbf{x}_{l-1} - \mu\mathbf{1}_{n_{l-1}}), \mathbf{\Sigma}_l - \mathbf{B}_l\mathbf{W}_l\mathbf{B}_l'),$$

where $\mathbf{B}_l = \boldsymbol{\Sigma}_l \mathbf{A}_l' \mathbf{W}_l^{-1}$ and $\mathbf{W}_l = \mathbf{A}_l \boldsymbol{\Sigma}_l \mathbf{A}_l' + \delta_l \mathbf{I}_{n_{l-1}}$. This concludes part (iii).
Therefore

$$p(\mathbf{x}_0, \ldots, \mathbf{x}_{L-1} | G_{L-1}) = p(\mathbf{x}_0 | G_{L-1}) \prod_{l=1}^{L-1} p(\mathbf{x}_l | G_{L-1}, \mathbf{x}_0, \ldots, \mathbf{x}_{l-1})$$

$$= p(\mathbf{x}_0 | I_0) \prod_{l=1}^{L-1} p(\mathbf{x}_l | I_l, \mathbf{x}_{l-1})$$

$$= N(\mathbf{x}_0 | \mu \mathbf{1}_{n_0}, \mathbf{Q}_0)$$

$$\prod_{l=1}^{L-1} N(\mathbf{x}_l | \mu \mathbf{1}_{n_l} + \mathbf{B}_l (\mathbf{x}_{l-1} - \mu \mathbf{1}_{n_{l-1}}), \boldsymbol{\Sigma}_l - \mathbf{B}_l \mathbf{W}_l \mathbf{B}_l').$$

This concludes part (i).

The proof of part (ii) is by induction. By Hypothesis (1), $p(\mathbf{x}_0 | G_{L-1}) = p(\mathbf{x}_0 | I_0) = N(\mathbf{x}_0 | \mu \mathbf{1}_{n_0}, \mathbf{Q}_0)$. Assume that $p(\mathbf{x}_{l-1} | G_{L-1}) = N(\mathbf{x}_{l-1} | \mu \mathbf{1}_{n_{l-1}}, \mathbf{Q}_{l-1})$. Then

$$p(\mathbf{x}_l | G_{L-1}) = \int p(\mathbf{x}_l | G_{L-1}, \mathbf{x}_{l-1}) p(\mathbf{x}_{l-1} | G_{L-1}) d\mathbf{x}_{l-1}$$

$$= \int p(\mathbf{x}_l | I_l, \mathbf{x}_{l-1}) p(\mathbf{x}_{l-1} | G_{L-1}) d\mathbf{x}_{l-1}$$

$$= \int N(\mathbf{x}_l | \mu \mathbf{1}_{n_l} + \mathbf{B}_l (\mathbf{x}_{l-1} - \mu \mathbf{1}_{n_{l-1}}), \boldsymbol{\Sigma}_l - \mathbf{B}_l \mathbf{W}_l \mathbf{B}_l')$$

$$N(\mathbf{x}_{l-1} | \mu \mathbf{1}_{n_{l-1}}, \mathbf{Q}_{l-1}) d\mathbf{x}_{l-1}$$

$$= N(\mathbf{x}_l | \mu \mathbf{1}_{n_l}, \mathbf{Q}_l),$$

where $\mathbf{Q}_l = \boldsymbol{\Sigma}_l - \mathbf{B}_l (\mathbf{W}_l - \mathbf{Q}_{l-1}) \mathbf{B}_l'$. This concludes part (ii).   □

In several types of problems, the direct computation of the matrices $\mathbf{B}_l$, $\mathbf{W}_l$, and $\mathbf{Q}_l$, $l = 1, \ldots, L - 1$, will not be necessary. Instead, it will be more useful to consider the following alternative representation of part (iii) of Theorem 10.2 for $l = 1, \ldots, L - 1$:

$$p(\mathbf{x}_l | G_l, \mathbf{x}_0, \ldots, \mathbf{x}_{l-1}) = \frac{N(\mathbf{x}_{l-1} | \mathbf{A}_l \mathbf{x}_l, \mathbf{U}_l) N(\mathbf{x}_l | \mu \mathbf{1}_{n_l}, \boldsymbol{\Sigma}_l)}{N(\mathbf{x}_{l-1} | \mu \mathbf{1}_{n_{l-1}}, \mathbf{A}_l \boldsymbol{\Sigma}_l \mathbf{A}_l' + \mathbf{U}_l)}. \quad (10.12)$$

This factorization allows efficient generation of realizations from the multiscale model in a cascade fashion from the coarsest to the finest levels of resolution.

10.3 The Multiscale Model as an Application of Jeffrey's Rule

This section presents an interpretation of the multiscale model of Ferreira et al. (2005) as an application of Jeffrey's rule.

Quoting Diaconis and Zabell (1982): "Jeffrey's rule for revising a probability P to a new probability P^* based on new probabilities $P^*(E_i)$ on a partition $\{E_i\}_{i=1,\ldots,n}$ is

$$P^*(A) = \sum P(A|E_i)P^*(E_i).$$

Jeffrey's rule is applicable if it is judged that $P^*(A|E_i) = P(A|E_i)$ for all A and i."

The construction of the multiscale model of Ferreira et al. (2005) can be interpreted as an application of Jeffrey's rule for revising probabilities in the following way. Initially, one would think that the finer level follows a Markov random field process, say, and the coarser level is derived from the finer level as a known function plus noise. In this way, the distribution of the coarser level is already determined, as well as the distribution of the finer level given the coarser level, denoted by $p(\mathbf{x}_l|I_l, \mathbf{x}_{l-1})$.

The hypothesis $p(\mathbf{x}_l|G_l, \mathbf{x}_0, \ldots, \mathbf{x}_{l-1}) = p(\mathbf{x}_l|G_{l-1}, I_l, \mathbf{x}_{l-1}) = p(\mathbf{x}_l|I_l, \mathbf{x}_{l-1})$ assumes the condition $P^*(A|E_i) = P(A|E_i)$, necessary for the application of Jeffrey's rule, as well as assuming Markovian resolution levels; that is, $(\mathbf{x}_0, \ldots, \mathbf{x}_{l-1})$ and $(\mathbf{x}_{l+1}, \ldots, \mathbf{x}_{L-1})$ are conditionally independent given $\mathbf{x}_l$. Thus, in the notation of Jeffrey's rule, $p(\mathbf{x}_l|G_l, \mathbf{x}_0, \ldots, \mathbf{x}_{l-1})$ corresponds to $P^*(A|E_i)$ and $p(\mathbf{x}_l|I_l, \mathbf{x}_{l-1})$ corresponds to $P(A|E_i)$.

It is fundamental to note that the assumption $p(\mathbf{x}_0, \ldots, \mathbf{x}_{l-1}|G_l) = p(\mathbf{x}_0, \ldots, \mathbf{x}_{l-1}|G_{l-1})$ in Section 10.2 means that the distribution of $(\mathbf{x}_0, \ldots, \mathbf{x}_{l-1})$ will be completely revised. In the notation of Jeffrey's rule, $p(\mathbf{x}_0, \ldots, \mathbf{x}_{l-1}|G_{l-1})$ corresponds to $P^*(E_i)$.

Finally, Jeffrey's rule can be used to recalculate the distribution of the finer level l, being equivalent to part (ii) of Theorem 10.2.

10.4 Didactic Example: Three-Level Model

This section illustrates the theory developed in Section 10.2 by the didactic construction of a three-level model. In particular, interest will be in the joint distribution of the three levels $\mathbf{x}_0$, $\mathbf{x}_1$, and $\mathbf{x}_2$. To make matters simpler, it is assumed that $\mu = 0$.

The initial information for each level is

$$p(\mathbf{x}_0|G_0) = p(\mathbf{x}_0|I_0) = N(\mathbf{x}_0|\mathbf{0}, \mathbf{\Sigma}_0) = N(\mathbf{x}_0|\mathbf{0}, \mathbf{Q}_0),$$

$$p(\mathbf{x}_1|I_1) = N(\mathbf{x}_1|\mathbf{0}, \mathbf{\Sigma}_1),$$

$$p(\mathbf{x}_2|I_2) = N(\mathbf{x}_2|\mathbf{0}, \mathbf{\Sigma}_2).$$

The link equation between $\mathbf{x}_0$ and $\mathbf{x}_1$ is

$$p(\mathbf{x}_0|I_1, \mathbf{x}_1) = N(\mathbf{x}_0|\mathbf{A}_1\mathbf{x}_1, \mathbf{U}_1).$$

Integrating out $\mathbf{x}_1$,

$$p(\mathbf{x}_0|I_1) = N(\mathbf{x}_0|\mathbf{0}, \mathbf{A}_1\mathbf{\Sigma}_1\mathbf{A}_1' + \mathbf{U}_1).$$

Moreover, by Bayes' Theorem,

$$\begin{aligned}
p(\mathbf{x}_1|I_1, \mathbf{x}_0) &= \frac{p(\mathbf{x}_1|I_1)p(\mathbf{x}_0|I_1, \mathbf{x}_1)}{p(\mathbf{x}_0|I_1)} \\
&= \frac{N(\mathbf{x}_1|\mathbf{0}, \mathbf{\Sigma}_1)N(\mathbf{x}_0|\mathbf{A}_1\mathbf{x}_1, \mathbf{U}_1)}{N(\mathbf{x}_0|\mathbf{0}, \mathbf{A}_1\mathbf{\Sigma}_1\mathbf{A}_1' + \mathbf{U}_1)}.
\end{aligned}$$

Thus, the joint distribution of $\mathbf{x}_0$ and $\mathbf{x}_1$ conditional on the information $G_1 = G_0 \cap I_1$ is

$$\begin{aligned}
p(\mathbf{x}_0, \mathbf{x}_1|G_1) &= p(\mathbf{x}_1|G_1, \mathbf{x}_0)p(\mathbf{x}_0|G_1) \\
&= p(\mathbf{x}_1|I_1, \mathbf{x}_0)p(\mathbf{x}_0|G_0) \\
&= \frac{N(\mathbf{x}_1|\mathbf{0}, \mathbf{\Sigma}_1)N(\mathbf{x}_0|\mathbf{A}_1\mathbf{x}_1, \mathbf{U}_1)}{N(\mathbf{x}_0|\mathbf{0}, \mathbf{A}_1\mathbf{\Sigma}_1\mathbf{A}_1' + \mathbf{U}_1)}N(\mathbf{x}_0|\mathbf{0}, \mathbf{Q}_0).
\end{aligned}$$

Consider now the resolution level $\mathbf{x}_2$. From the link equation (10.11),

$$p(\mathbf{x}_1|I_2, \mathbf{x}_2) = N(\mathbf{x}_1|\mathbf{A}_2\mathbf{x}_2, \mathbf{U}_2).$$

Thus, by Bayes' Theorem,

$$\begin{aligned}
p(\mathbf{x}_2|I_2, \mathbf{x}_1) &= \frac{p(\mathbf{x}_2|I_2)p(\mathbf{x}_1|I_2, \mathbf{x}_2)}{p(\mathbf{x}_1|I_2)} \\
&= \frac{N(\mathbf{x}_2|\mathbf{0}, \mathbf{\Sigma}_2)N(\mathbf{x}_1|\mathbf{A}_2\mathbf{x}_2, \mathbf{U}_2)}{N(\mathbf{x}_1|\mathbf{0}, \mathbf{A}_2\mathbf{\Sigma}_2\mathbf{A}_2' + \mathbf{U}_2)}.
\end{aligned}$$

Therefore, the joint distribution of $\mathbf{x}_0$, $\mathbf{x}_1$, and $\mathbf{x}_2$ conditional on the information $G_2 = G_1 \cap I_2$ is

$$\begin{aligned}
p(\mathbf{x}_0, \mathbf{x}_1, \mathbf{x}_2|G_2) &= p(\mathbf{x}_2|G_2, \mathbf{x}_0, \mathbf{x}_1)p(\mathbf{x}_0, \mathbf{x}_1|G_2) \\
&= p(\mathbf{x}_2|I_2, \mathbf{x}_1)p(\mathbf{x}_0, \mathbf{x}_1|G_1) \\
&= \frac{N(\mathbf{x}_2|\mathbf{0}, \mathbf{\Sigma}_2)N(\mathbf{x}_1|\mathbf{A}_2\mathbf{x}_2, \mathbf{U}_2)}{N(\mathbf{x}_1|\mathbf{0}, \mathbf{A}_2\mathbf{\Sigma}_2\mathbf{A}_2' + \mathbf{U}_2)} \\
&\quad \times \frac{N(\mathbf{x}_1|\mathbf{0}, \mathbf{\Sigma}_1)N(\mathbf{x}_0|\mathbf{A}_1\mathbf{x}_1, \mathbf{U}_1)}{N(\mathbf{x}_0|\mathbf{0}, \mathbf{A}_1\mathbf{\Sigma}_1\mathbf{A}_1' + \mathbf{U}_1)} \\
&\quad \times N(\mathbf{x}_0|\mathbf{0}, \mathbf{Q}_0).
\end{aligned} \qquad (10.13)$$

This three-level model is used in Chapter 16 for modeling 1-D permeability fields.

10.5 Posterior Simulation

The posterior distribution is explored by using a Markov chain Monte Carlo procedure (Gamerman and Lopes 2006) that generates a sample from the joint posterior density of the unknown quantities of the model. From this sample, posterior summaries such as means, medians, variances, and credible intervals can be computed through Monte Carlo integration. Section 10.5.1 describes the simulation of multiscale random fields, while Section 10.5.2 describes the simulation of the related parameters.

10.5.1 Simulation of Multiscale Random Fields

Simulation of multiscale random fields is necessary for the inference process when part of the field is observed and part is not or when these models are used as priors for hidden processes. When the relationship between the hidden process and the observations is nonlinear and nonlocal, as in the inverse problem application presented in Chapter 16, simulation of the multiscale random field is performed using MCMC-based iterative methods. Here the focus is on iterative simulation methods.

As the coarsest level $\mathbf{x}_0$ follows a Markov random field a priori, its simulation is straightforward. In order to explore the joint distribution of the random field sites, single-site Metropolis steps are used. More specifically, the proposal for x_{0k} is simulated from

$$x_{0k}^* \sim N\left(x_{0k}^{(old)}, C_{0k}/\phi_{0k}\right),$$

where $C_{0k} = \{\tau_0[\alpha_0 + h_{0k}]\}^{-1}$ and ϕ_{0k} is adjusted to yield good acceptance rates. The proposal is accepted with the usual Metropolis step acceptance probability (Gamerman and Lopes, 2006).

The simulation of the lth level ($l = 1, \ldots, L-1$) is performed by blocks using a checkerboard pattern, with each block corresponding to one site of the immediate coarser level. Let $Bl_{l-1,j}$ and $Re_{l-1,j}$ be the black and white sites, respectively, of level l corresponding to site $(l-1, j)$. As simulation of white and black sites is analogous, here the latter is presented.

The prior full conditional density for black sites at level l corresponding to site $(l-1, k)$ is

$$p(\mathbf{x}_{Bl_{l-1,k}}|G_{L-1}, \mathbf{x}_{\sim Bl_{l-1,k}}, \mathbf{x}_{l-1}) \propto p(\mathbf{x}_l|I_l, \mathbf{x}_{l-1})$$

$$= \frac{p(\mathbf{x}_{l-1}|I_l, \mathbf{x}_l)p(\mathbf{x}_l|I_l)}{p(\mathbf{x}_{l-1}|I_l)}$$

$$\propto p(x_{l-1,k}|I_l, \mathbf{x}_{D_{l-1,k}}) \prod_{j \in Bl_{l-1,k}} p(x_{lj}|I_l, \mathbf{x}_{N_{lj}})$$

$$\propto \exp\left\{-\frac{1}{2\delta_l}\left(x_{l-1,k} - m_l^{-1}\sum_{j\in D_{l-1,k}} x_{lj}\right)^2\right\}$$

$$\prod_{j\in Bl_{l-1,k}} \exp\left\{-\frac{l}{2C_{lj}}(x_{lj} - a_{lj})^2\right\}, \tag{10.14}$$

where

$$C_{lj} = [\tau_l(\alpha_l + h_{lj})]^{-1}$$

and

$$a_{lj} = \tau_l C_{lj}\left(\alpha_l\mu + \sum_{i\in N_{lj}} g_{lij}x_{li}\right).$$

The joint proposal for $\mathbf{x}_{Bl_{l-1,k}}$ is simulated as follows. The proposal for each site in $Bl_{l-1,k}$ is simulated independently from

$$x_{lk}^* \sim N\left(x_{lk}^{(old)}, C_{lk}/\phi_{lk}\right),$$

where ϕ_{ik} is a tuning parameter adjusted to yield good acceptance rates. The joint proposal for $\mathbf{x}_{Bl_{l-1,k}}$ is accepted with Metropolis acceptance probability.

10.5.2 Parameter Updates

Here the simulation of the parameters and their prior specification is considered. The parameters are assumed independent a priori with joint prior density

$$p(\mu)\prod_{l=0}^{L-1} p(\alpha_l)p(\tau_l)\prod_{l=1}^{L-1} p(\delta_l).$$

More specifically,

$$p(\mu) = N(\mu|m_\mu, s_\mu^2),$$

$$p(\delta_l) = Ga(\delta_l|0.5n_{\delta_l}, 0.5n_{\delta_l}s_{\delta_l}^2),$$

$$p(\tau_l) = Ga(\tau_l|0.5n_{\tau_l}, 0.5n_{\tau_l}s_{\tau_l}^2),$$

$$p(\alpha_l) \propto \left[\sum_{k=1}^{n_l-1}(\lambda_{lk} + \alpha_l)^{-2} - (n_l - 1)^{-1}\left\{\sum_{k=1}^{n_l-1}(\lambda_{lk} + \alpha_l)^{-1}\right\}^2\right]^{0.5},$$

where this last expression is the reference prior derived by Ferreira and de Oliveira (2007) for the spatial parameter of a PMRF.

The mean level μ is simulated with a Metropolis step. The acceptance probability is computed with the help of Equation (10.12).

The simulation of δ_l, $l = 1,\ldots,L - 1$, is performed with a Metropolis-Hastings step,

$$\delta_l^* \sim U\left(\delta_l^{(old)}/\phi_\delta, \delta_l^{(old)}\phi_\delta\right),$$

where $\phi_\delta > 1$ is a tuning parameter adjusted to yield good acceptance rates.

The parameters α_l and τ_l, $l = 0, \ldots, L-1$, are strongly correlated a posteriori. In order to overcome the associated difficulties, α_l and τ_l are jointly simulated using individual Metropolis-Hastings steps,

$$\alpha_l^* \sim U\left(\alpha_l^{(old)}/\phi_\alpha, \alpha_l^{(old)}\phi_\alpha\right)$$

and

$$\tau_l^* \sim U\left(\tau_l^{(old)}/\phi_\tau, \tau_l^{(old)}\phi_\tau\right).$$

The joint proposal (α_l^*, τ_l^*) is accepted with the appropriate Metropolis-Hastings probability.

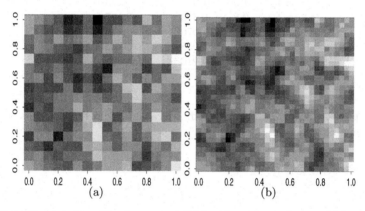

Fig. 10.3. Simulated multiscale random field with parameters $\mu = 15.0$, $\tau_0 = 1.0$, $\alpha_0 = 0.5$, $\tau_1 = 8.0$, $\alpha_1 = 1.0$, and $\delta_1 = 0.0016$. (a) Coarse level; (b) fine level.

10.6 A Simulated Example

Figure 10.3 presents a simulated multiscale random field with two-levels of resolution: a 16×16 coarse level and a 32×32 fine level. Simulation of this multiscale random field was performed as discussed in Section 10.5.1. The parameter values used to simulate this multiscale random field were $\mu = 15.0$, $\tau_0 = 1.0$, $\alpha_0 = 0.5$, $\tau_1 = 8.0$, $\alpha_1 = 1.0$, and $\delta_1 = 0.0016$. The values of α_0 and α_1 imply a strong spatial dependence at both the coarse level and the fine level. The value of τ_0 leads to a reasonable variability at the coarse level, while the value of τ_1 leads to less difference between neighbors at the fine level. The value of δ_1 imposes a strong constraint on the fine level, causing it to closely follow the realized coarse level.

Note that the fine level has a local behavior similar to that of Markov random fields yet retains the coarse-level-induced global behavior. This global behavior is analogous to the long memory type of behavior observed in multiscale time series models (Ferreira et al., 2006).

The parameters were estimated using the MCMC-based procedure presented in Section 10.5. Vague priors were used with $m_\mu = 0$, $s_\mu^2 = 10^4$, and $n_{\delta_1} = n_{\delta_1} s_{\delta_1}^2 = n_{\tau_0} = n_{\tau_0} s_{\tau_0}^2 = n_{\tau_1} = n_{\tau_1} s_{\tau_1}^2 = 10^{-3}$. Table 10.1 presents some posterior summaries showing that the estimation procedure works very well.

Table 10.1. Simulated multiscale random field – Posterior summaries.

	True value	Mean	Standard deviation
μ	15.0	15.024	0.053
τ_0	1.0	1.066	0.123
α_0	0.5	0.308	0.197
τ_1	8.0	7.970	0.446
α_1	1.0	0.955	0.123
δ_1	0.0016	0.00161	0.00009

Chapter 16 discusses the use of multiscale random field models as priors for permeability fields.

11

Multiscale Time Series

11.1 Introduction

This chapter presents multiscale time series models that have been introduced by Ferreira et al. (2006). These models couple standard linear models at different time scales with linear link equations between scales. As in the multiscale random field models presented in Chapter 10, Jeffrey's rule of conditioning is used to ensure that the processes at the different levels are compatible. This allows consistent modeling of time series at different levels of resolution (e.g., daily or monthly aggregates of financial or meteorological data) and also allows coherent combination of information across time scales.

Concerns about the temporal scales of resolution of time series date back at least to Working (1960), who studied the effect of aggregation of random-walk processes. Since then, several authors have considered the use of data sampled at a coarser level of resolution than the original time scale on which the stochastic model is defined; for example, Telser (1967), Amemiya and Wu (1972), Palm and Nijman (1984), Drost and Nijman (1993), Schmidt and Gamerman (1997), Hwang (2000), and Bollerslev and Wright (2000). Rather than using models defined at a fine resolution and aggregated to coarser resolutions, Ferreira et al. (2006) have proposed a new class of multiscale time series models built from coarse to fine scales of resolution.

The coarse to fine construction results in a class of models for time series that has the capacity to emulate long memory processes. Formally, a long memory process has autocorrelation function $\rho(k) = ck^{2d-1}$, where $c \neq 0$ and $0 < d < 0.5$. That is, the decay of the autocorrelation function is geometric, instead of exponential as for autoregressive processes. Thus, for a long memory process, as $\sum_{k=1}^{\infty} \rho(k) = \infty$, the spectral function approaches infinity in the limit when the frequency goes to zero. For references on long memory processes, see Beran (1994) and Brockwell and Davis (1991). The multiscale time series models of Ferreira et al. (2006) are built in a cascade way from coarse to fine levels of resolution. The coarser levels of resolution will determine the large time scale behavior of the series, and the finer levels will determine the high-frequency behavior. Analogous to the construction of long memory processes through the use of the inverse discrete wavelet transform (Wornell, 1990), if an infinite number of resolutions are used, then the resulting process at the finest level is a long memory process. Usually a small number of resolution levels, such as two or three, is enough to model the time series of interest and to induce an approximate long memory process at the finest time-resolution level.

Long memory is only one possible behavior of the models proposed by Ferreira et al. (2006). These models are adequate for naturally multiscale processes; that is, processes that exist and have different dynamics at different resolution levels. Moreover, with a very parsimonious parameterization based on a few parameters for each resolution, the induced process at the finest resolution level exhibits a variety of autocorrelation structures. This is useful when the process is naturally multiscale but data are observed only at the finest scale; in this case, the model is used to induce a particular process at this finest scale, resulting in what Ferreira et al. (2006) call Hidden Resolution Models (HRMs). As there is a progressive correlated stochastic process at each scale, the resulting multiscale class of models can be used in extrapolation and forecasting.

Another useful feature of the multiscale time series models of Ferreira et al. (2006) is the ability to combine information across levels of resolution. These models are proper joint probability models for the process at the different levels of resolution. Thus, in the case of data observed at different scales, the models can be used to coherently integrate the information from the different scales. This is accomplished with standard probability calculus operations such as marginalization and conditioning.

The remainder of this chapter is organized as follows. Section 11.2.1 presents the general framework for multiscale time series models (MSTSMs) and HRMs with one coarse level and one fine level. Section 11.2.2 presents an example of an MSTSM and the corresponding HRM with autoregressive processes of order 1 as building blocks. Section 11.2.3 shows the derivation, for the general case, of the implied distribution at the fine level; that is, the Hidden Resolution Model. Section 11.3 presents several properties of HRMs. Section 11.4 describes extensions to incorporate periodicities into the multiscale framework relevant for certain applications. Section 11.5 discusses issues of inference and prediction. Three examples are presented in Section 11.6.

11.2 Model Construction

11.2.1 General Framework

Begin with a univariate time series x_t, $(t = 1, 2, \ldots)$, following a specified model. Generally, write $\mathbf{x}_{1:n_x} = (x_1, \ldots, x_{n_x})$ for any integer $n_x > 0$, and denote by $p(\mathbf{x}_{1:n_x})$ the density of the joint distribution of $\mathbf{x}_{1:n_x}$. For example, the x_t could follow a standard linear stationary time series model, such as an AR(1) model, that is completely specified; in that case, $p(\mathbf{x}_{1:n_x})$ is the implied stationary distribution for the set of n_x values. In other examples, $p(\mathbf{x}_{1:n_x})$ could be a posterior predictive distribution arising from a dynamic model conditioned on past observations (as in West and Harrison, 1997).

For a specified positive integer m, define the process y_s indexed on $s = 1, 2, \ldots$ by $y_s = m^{-1} \sum_{i=1}^{m} x_{(s-1)m+i} + u_s$, where the u_s, ($s = 1, 2, \ldots$), are mutually uncorrelated zero-mean, normally distributed noise terms with $u_s \sim N(u_s | 0, \tau)$ for some between-level variance τ. The y values are the averages of nonoverlapping groups of m consecutive x values, such as weekly averages ($m = 7$) of daily data or annual averages ($m = 12$) of monthly data, subject to the addition of noise or error terms u. Refer to m as the coarsening window, x_t as the *fine-level* process, and y_s as the *coarse-level* aggregate process. Then, the marginal distribution for $\mathbf{y}_{1:n_y}$ is computed as $p(\mathbf{y}_{1:n_y}) = \int p(\mathbf{y}_{1:n_y}|\mathbf{x}_{1:n_x})p(\mathbf{x}_{1:n_x})d\mathbf{x}_{1:n_x}$. The implied distribution of the process at the coarse level is usually less tractable and more complex than at the fine level.

The basic idea underlying the multiscale construction, analogous to that of Chapter 10, is to impose a simple process at the coarse level. This is interpreted as a new piece of information G received after $p(\mathbf{x}_{1:n_x})$ and $p(\mathbf{y}_{1:n_y}|\mathbf{x}_{1:n_x})$ are defined and that supersedes the prior information on which $p(\mathbf{x}_{1:n_x})$ is based. To be specific, suppose that the additional information G relevant to y has as a consequence the revision of the distribution of $\mathbf{y}_{1:n_y}$ to $q(\mathbf{y}_{1:n_y})$. That is equivalent to the existence of an implicit likelihood function for $\mathbf{y}_{1:n_y}$ denoted by $l(\mathbf{y}_{1:n_y}|G)$ and computed as $l(\mathbf{y}_{1:n_y}|G) \propto q(\mathbf{y}_{1:n_y})/p(\mathbf{y}_{1:n_y})$.

It is extremely important to note that $p(\mathbf{x}_{1:n_x})$, $q(\mathbf{y}_{1:n_y})$, and $p(\mathbf{y}_{1:n_y}|\mathbf{x}_{1:n_x})$ may be inconsistent and generally will be. Consistency is restored by the use of Jeffrey's rule of conditioning to revise the fine-scale model $p(\mathbf{x}_{1:n_x})$. In this way, the models are made consistent across the different levels of resolution. Diaconis and Zabell (1982) discuss the use and roles of Jeffrey's rule in Bayesian analysis. Loschi et al. (2002) use Jeffrey's rule for inference in the context of financial time series. Good references on Jeffrey's rule of conditioning are Jeffrey (1992), Diaconis and Zabell (1982), and Shafer (1981). Jeffrey's rule of conditioning updates the fine-level distribution by the formula $q(\mathbf{x}_{1:n_x}) = \int p(\mathbf{x}_{1:n_x}|\mathbf{y}_{1:n_y})q(\mathbf{y}_{1:n_y})d\mathbf{y}_{1:n_y}$. If the two levels of resolution $\mathbf{x}_{1:n_x}$ and $\mathbf{y}_{1:n_y}$ are observable, then the Multiscale Time Series Model (MSTSM) is $p(\mathbf{x}_{1:n_x}|\mathbf{y}_{1:n_y})q(\mathbf{y}_{1:n_y})$. Conversely, if the coarse level $\mathbf{y}_{1:n_y}$ is an unobservable latent process, then the implied process at the fine level $q(\mathbf{x}_{1:n_x})$ is called a Hidden Resolution Model (HRM).

Section 11.2.2 presents an example using autoregressive processes of order 1 as building blocks for the construction of MSTSMs and HRMs.

11.2.2 Example – AR(1) Building Blocks

Suppose that the initial fine-level model is a standard stationary linear AR(1) model

$$x_t = \phi_x x_{t-1} + \varepsilon_t, \tag{11.1}$$

where ε_t, $(t = 1, 2, \ldots)$, is a sequence of mutually uncorrelated zero-mean, normally distributed innovations with $\varepsilon_t \sim N(0, \sigma_x^2)$ for some variance σ_x^2. Then, for all $n_x > 0$, $p(\mathbf{x}_{1:n_x})$ is the implied n_x-dimensional stationary distribution

$$p(\mathbf{x}_{1:n_x}) = N(\mathbf{x}_{1:n_x} | \mathbf{0}, \mathbf{V}_x), \tag{11.2}$$

where $\mathbf{0}$ is the vector of n_x zeros and $\mathbf{V}_x$ is the n_x-square variance matrix with i, j element $\sigma_x^2 \phi_x^{|i-j|} / (1 - \phi_x^2)$.

Suppose that the link between levels is described simply via the densities $p(\mathbf{y}_{1:n_y} | \mathbf{x}_{1:n_x})$ (for all n_y) implied by assuming that the y_s are conditionally independent and normally distributed with means $m^{-1} \sum_{i=1}^{m} x_{(s-1)m+i}$ and constant variance τ.

Fixing $n_x = n_y m$, define the $n_y \times n_x$ matrix $\mathbf{A}$ such that $E[\mathbf{y}_{1:n_y}] = \mathbf{A} \mathbf{x}_{1:n_x}$. Thus $\mathbf{A}$ is a sparse matrix whose nonzero elements are all $1/m$; in row i, the nonzero elements are those in columns $(i - 1)m + 1$ to im. Then, the link equation is defined as

$$p(\mathbf{y}_{1:n_y} | \mathbf{x}_{1:n_x}) = \prod_{s=1}^{n_y} N\left(y_s \middle| m^{-1} \sum_{i=1}^{m} x_{(s-1)m+i}, \tau\right) = N(\mathbf{y}_{1:n_y} | \mathbf{A} \mathbf{x}_{1:n_x}, \mathbf{U}), \tag{11.3}$$

where $\mathbf{U} = \tau \mathbf{I}$, with $\mathbf{I}$ as the n_y-square identity matrix and *between-levels* variance τ. It is useful to parameterize τ as a function of $\mathbf{A} \mathbf{V}_x \mathbf{A}'$, and the parameterization $\tau = \lambda (\mathbf{A} \mathbf{V}_x \mathbf{A}')_{11}$ works well. As the parameter λ has a natural interpretation in terms of the relative increase in uncertainty at the coarse level due to the lack of agreement with the fine level, it is much easier to establish a prior for λ than for τ.

Suppose that additional information about the coarse level $\mathbf{y}_{1:n_y}$ is now received such that the revised model $q(\mathbf{y}_{1:n_y})$ is a simple standard stationary linear AR(1) model

$$y_s = \phi_y y_{s-1} + \eta_s, \tag{11.4}$$

where η_s, $(s = 1, 2, \ldots,)$, is a sequence of mutually uncorrelated zero-mean, normally distributed innovations with variance σ_y^2. Then, for all $n_y > 0$, $q(\mathbf{y}_{1:n_y})$ is the implied n_y-dimensional stationary distribution

$$q(\mathbf{y}_{1:n_y}) = N(\mathbf{y}_{1:n_y} | \mathbf{0}, \mathbf{Q}_y), \tag{11.5}$$

where $\mathbf{0}$ is the vector of n_y zeros and $\mathbf{Q}_y$ is the n_y-square variance matrix with i, j element $\sigma_y^2 \phi_y^{|i-j|} / (1 - \phi_y^2)$.

Obviously, the densities $p(\mathbf{y}_{1:n_y})$ and $q(\mathbf{y}_{1:n_y})$ are not compatible. However, viewing the information G as superseding the information on which the model

$p(\mathbf{x}_{1:n_x})$ is based, Equation (11.5) must be adopted and $p(\mathbf{x}_{1:n_x})$ has to be updated accordingly. This is done using Jeffrey's rule of conditioning to revise the distribution of $\mathbf{x}_{1:n_x}$. The application of Jeffrey's rule assumes that the revised conditional distribution of $\mathbf{x}_{1:n_x}$ given $\mathbf{y}_{1:n_y}$, denoted by $q(\mathbf{x}_{1:n_x}|\mathbf{y}_{1:n_y})$, is equal to the conditional distribution of $\mathbf{x}_{1:n_x}$ given $\mathbf{y}_{1:n_y}$ implied by Equations (11.2) and (11.3), denoted by $p(\mathbf{x}_{1:n_x}|\mathbf{y}_{1:n_y})$. As discussed in Chapter 10, this assumption means that given $\mathbf{y}_{1:n_y}$, $\mathbf{x}_{1:n_x}$ is independent of the new information that led to the revision of beliefs about $\mathbf{y}_{1:n_y}$.

Section 11.2.3 shows that the resulting distribution $q(\mathbf{x}_{1:n_x})$, the Hidden Resolution Model, is also zero-mean normal, $q(\mathbf{x}_{1:n_x}) = N(\mathbf{x}_{1:n_x}|\mathbf{0}, \mathbf{Q}_x)$, with covariance matrix $\mathbf{Q}_x = \mathbf{V}_x - \mathbf{B}(\mathbf{W} - \mathbf{Q}_y)\mathbf{B}'$, where $\mathbf{W} = \mathbf{A}\mathbf{V}_x\mathbf{A}' + \mathbf{U}$ and $\mathbf{B} = \mathbf{V}_x\mathbf{A}'\mathbf{W}^{-1}$.

11.2.3 General Case – Implied Fine-Level Distribution

This section derives the revised distribution at the fine level; that is, the Hidden Resolution Model. This development is analogous to that of Section 10.1 for the case $\mu = 0$, and it is presented here for completeness.

Start by using Bayes' Theorem in order to obtain the conditional distribution of $\mathbf{x}_{1:n_x}$ given $\mathbf{y}_{1:n_y}$:

$$p(\mathbf{x}_{1:n_x}|\mathbf{y}_{1:n_y}) \propto p(\mathbf{x}_{1:n_x})p(\mathbf{y}_{1:n_y}|\mathbf{x}_{1:n_x})$$
$$= N(\mathbf{x}_{1:n_x}|\mathbf{0}, \mathbf{V}_x)\, N(\mathbf{y}_{1:n_y}|\mathbf{A}\mathbf{x}_{1:n_x}, \mathbf{U}).$$

Therefore, by linear regression, $\mathbf{x}_{1:n_x}|\mathbf{y}_{1:n_y} \sim N(\mathbf{B}\mathbf{y}_{1:n_y}, \mathbf{V}_x - \mathbf{B}\mathbf{W}\mathbf{B}')$, where $\mathbf{B} = \mathbf{V}_x\mathbf{A}'\mathbf{W}^{-1}$ and $\mathbf{W} = \mathbf{A}\mathbf{V}_x\mathbf{A}' + \mathbf{U}$. As the distribution $p(\mathbf{x}_{1:n_x})$ has mean zero, the expected value of $\mathbf{x}_{1:n_x}$ is shrunk from the corresponding value of $\mathbf{y}_{1:n_y}$ toward zero.

So, as $q > 0$ and $p > 0$, the generalized Jeffrey's rule (Diaconis and Zabell, 1982) can be used to derive the Hidden Resolution Model:

$$q(\mathbf{x}_{1:n_x}) = \int p(\mathbf{x}_{1:n_x}|\mathbf{y}_{1:n_y})q(\mathbf{y}_{1:n_y})d\mathbf{y}_{1:n_y}$$
$$\propto \int N(\mathbf{x}_{1:n_x}|\mathbf{B}\mathbf{y}_{1:n_y}, \mathbf{V}_x - \mathbf{B}\mathbf{W}\mathbf{B}')$$
$$N(\mathbf{y}_{1:n_y}|\mathbf{0}, \mathbf{Q}_y)d\mathbf{y}_{1:n_y}. \tag{11.6}$$

Therefore, the resulting distribution $q(\mathbf{x}_{1:n_x})$, the Hidden Resolution Model, is also zero-mean normal,

$$q(\mathbf{x}_{1:n_x}) = N(\mathbf{x}_{1:n_x}|\mathbf{0}, \mathbf{Q}_x), \tag{11.7}$$

with covariance matrix

$$\mathbf{Q}_x = \mathbf{V}_x - \mathbf{B}(\mathbf{W} - \mathbf{Q}_y)\mathbf{B}', \tag{11.8}$$

where $\mathbf{W} = \mathbf{A}\mathbf{V}_x\mathbf{A}' + \mathbf{U}$ and $\mathbf{B} = \mathbf{V}_x\mathbf{A}'\mathbf{W}^{-1}$.

11.2.4 Example – MA(1) Building Blocks

Assume that the initial fine-level model is an MA(1) process,

$$x_t = \varepsilon_t + \theta_x \varepsilon_{t-1},$$

where the $\{\varepsilon_t\}$ are i.i.d. $N(0, \sigma_\varepsilon^2)$. Then, $\mathbf{x}_{1:n_x} \sim N(\mathbf{0}, \mathbf{V}_x)$, where $\mathbf{V}_x$ is a sparse matrix with all but three diagonals equal to zero. The main diagonal elements are equal to $(1 + \theta_x^2)\sigma_\varepsilon^2$, and the elements of the subdiagonals and superdiagonals are equal to $\theta_x \sigma_\varepsilon^2$.

Consider the link equation

$$p(\mathbf{y}_{1:n_y} | \mathbf{x}_{1:n_x}) = \prod_{s=1}^{n_y} N\left(y_s \middle| m^{-1} \sum_{i=1}^{m} x_{(s-1)m+i}, \tau\right) = N(\mathbf{y}_{1:n_y} | \mathbf{A}\mathbf{x}_{1:n_x}, \mathbf{U}),$$

(11.9)

where $\mathbf{U} = \tau\mathbf{I}$ with $\mathbf{I}$ as the n_y-square identity matrix and *between-levels* variance τ.

Then, $p(\mathbf{y}_{1:n_y}) = N(\mathbf{y}_{1:n_y} | \mathbf{0}, \mathbf{A}\mathbf{V}_x\mathbf{A}' + \mathbf{U})$. Note that as $\mathbf{V}_x$ is a sparse matrix and $\mathbf{U}$ is a diagonal matrix, then $\mathbf{W} = \mathbf{A}\mathbf{V}_x\mathbf{A}' + \mathbf{U}$ is a sparse matrix, a fact that can be used to accelerate computations when using MA(1) building blocks.

In order to determine the implied process at the coarse level, rewrite y_s as a function of the ε's:

$$y_s = m^{-1} \sum_{i=1}^{m} x_{(s-1)m+i} + u_s$$

$$= m^{-1} \sum_{i=1}^{m} (\varepsilon_{(s-1)m+i} + \theta_x \varepsilon_{(s-1)m+i-1}) + u_s$$

$$= m^{-1} \sum_{i=1}^{m} \varepsilon_{(s-1)m+i} + m^{-1}\theta_x \sum_{i=0}^{m-1} \varepsilon_{(s-1)m+i} + u_s$$

$$= m^{-1}(1 + \theta_x) \sum_{i=1}^{m-1} \varepsilon_{(s-1)m+i} + m^{-1}(\varepsilon_{sm} + \theta_x \varepsilon_{(s-1)m}) + u_s.$$

Define $\tilde{u}_s = u_s + m^{-1}(1 + \theta_x) \sum_{i=1}^{m-1} \varepsilon_{(s-1)m+i}$ and $\tilde{\varepsilon}_s = m^{-1}\varepsilon_{sm}$. Then $V(\tilde{u}_s) = \tau + m^{-2}(m-1)(1 + \theta_x)^2\sigma_\varepsilon^2$, $\text{Cov}(\tilde{u}_s, \tilde{u}_{s-1}) = 0$, $V(\tilde{\varepsilon}_s) = m^{-2}\sigma_\varepsilon^2$, and $\text{Cov}(\tilde{\varepsilon}_s, \tilde{\varepsilon}_{s-1}) = 0$. Thus, it is possible to rewrite y_s as

$$y_s = \tilde{u}_s + \tilde{\varepsilon}_s + \theta_x \tilde{\varepsilon}_{s-1}.$$

(11.10)

Then,

$$V(y_s) = \tau + \frac{(m-1)(1 + \theta_x)^2}{m^2}\sigma_\varepsilon^2 + m^{-2}(1 + \theta_x^2)\sigma_\varepsilon^2,$$

(11.11)

$$\text{Cov}(y_s, y_{s-1}) = E(y_s y_{s-1}) = \theta_x E(\tilde{\varepsilon}_{s-1}^2) = m^{-2} \theta_x \sigma_\varepsilon^2, \qquad (11.12)$$

and $\text{Cov}(y_s, y_{s-k}) = 0, \forall k \geq 2$. Therefore, y_s follows an MA(1) process with parameters that depend on θ_x and σ_ε^2.

Now, revise the distribution of $\mathbf{y}_{1:n_y}$ to an MA(1) process with parameters θ_y and σ_y^2. Using Jeffrey's rule to revise the distribution of $\mathbf{x}_{1:n_x}$, one obtains the Hidden Resolution Model of $\mathbf{x}_{1:n_x}$ given by the general Equations (11.7) and (11.8). Section 11.3.2 studies in some detail the behavior of autocorrelation functions of HRMs when the building blocks are MA(1) processes.

11.2.5 HRMs with ARMA Building Blocks

Sections 11.2.2 and 11.2.4 presented in detail the construction of Hidden Resolution Models with autoregressive and moving average processes of order 1 as building blocks. This section briefly discusses the implications of using autoregressive moving average (ARMA) processes as building blocks.

From a general viewpoint, the construction of MSTSMs and HRMs has three ingredients: the initial process at the fine level, the link equation, and the revised process at the coarse level. With these three ingredients, Jeffrey's rule can be used to derive the Hidden Resolution Model. Therefore, the construction put forward in this chapter is quite general and can be used with any well-behaved types of processes as building blocks. In particular, natural choices for building blocks are stationary and invertible ARMA processes.

The main implications of the use of ARMA processes as building blocks are related to the properties of the resulting HRM and to the particular methods necessary to estimate the parameters of the model and perform forecasts for future observations. Section 11.3.1 studies the limiting behavior of the covariance structure of HRMs for fairly general building blocks including ARMA processes. Properties of the updated fine-level process with AR(1) only, MA(1) only, and combinations of AR(1) and MA(1) building blocks are discussed in Section 11.3. These building blocks provide a very rich collection of Hidden Resolution Models, as shown in Figures 11.3 to 11.18. Properties of Hidden Resolution Models constructed with more general ARMA building blocks can be studied along the same lines as in Section 11.3 and will not be considered here. With respect to estimation and forecasting, when using general ARMA building blocks, the general idea is the same as the idea presented in Section 11.5 for the two-level model with AR(1) building blocks: the factorization of $p(\mathbf{x}_{1:n_x}|\mathbf{y}_{1:n_y})$ and the simple form of $q(\mathbf{y}_{1:n_y})$ are used to speed up the computations.

11.3 Properties of Hidden Resolution Models

Here we discuss some interesting properties of Hidden Resolution Models by analyzing their autocorrelation functions. In Section 11.3.1, limiting cases of the Hidden Resolution Model are considered, and in Section 11.3.2 autocorrelation functions for several Hidden Resolution Models are discussed. As a by-product of the study of the autocorrelation functions, the parameters of Hidden Resolution Models are given suitable interpretations.

11.3.1 Limiting Behavior

Some properties of Hidden Resolution Models for limiting cases are considered here. The focus here is on the behavior of some interesting functions of the covariance matrix $\mathbf{Q}_x$, such as the autocorrelation function and $\mathbf{A}\mathbf{Q}_x\mathbf{A}'$. All the theorems in this section are valid for Hidden Resolution Models constructed with the link equation (11.3) and stationary and invertible processes as building blocks.

Theorem 11.1 states that when $\tau \to 0$, the covariance structure of a coarsened version of an HRM process converges to the covariance structure of the hidden coarse-level process.

Theorem 11.1. $\lim_{\tau \to 0} \mathbf{A}\mathbf{Q}_x\mathbf{A}' = \mathbf{Q}_y$.

Proof.

$$
\begin{aligned}
\lim_{\tau \to 0} \mathbf{A}\mathbf{Q}_x\mathbf{A}' &= \mathbf{A}\big(\lim_{\tau \to 0}\mathbf{Q}_x\big)\mathbf{A}' \\
&= \mathbf{A}[\mathbf{V}_x - \mathbf{V}_x\mathbf{A}'(\mathbf{A}\mathbf{V}_x\mathbf{A}')^{-1}\mathbf{A}\mathbf{V}_x + \mathbf{V}_x\mathbf{A}'(\mathbf{A}\mathbf{V}_x\mathbf{A}')^{-1} \\
&\quad \mathbf{Q}_y(\mathbf{A}\mathbf{V}_x\mathbf{A}')^{-1}\mathbf{A}\mathbf{V}_x]\mathbf{A}' \\
&= \mathbf{A}\mathbf{V}_x\mathbf{A}' - \mathbf{A}\mathbf{V}_x\mathbf{A}'(\mathbf{A}\mathbf{V}_x\mathbf{A}')^{-1}\mathbf{A}\mathbf{V}_x\mathbf{A}' \\
&\quad +\mathbf{A}\mathbf{V}_x\mathbf{A}'(\mathbf{A}\mathbf{V}_x\mathbf{A}')^{-1}\mathbf{Q}_y(\mathbf{A}\mathbf{V}_x\mathbf{A}')^{-1}\mathbf{A}\mathbf{V}_x\mathbf{A}' \\
&= \mathbf{Q}_y.
\end{aligned}
$$

$\square$

Theorem 11.2 states that when $\tau \to \infty$, the covariance structure of the HRM process converges to the original covariance structure.

Theorem 11.2. $\lim_{\tau \to \infty} \mathbf{Q}_x = \mathbf{V}_x$.

Proof. Note that $\lim_{\tau \to \infty} \mathbf{W}^{-1} = \lim_{\tau \to \infty}(\mathbf{A}\mathbf{V}_x\mathbf{A}' + \tau\mathbf{I})^{-1} = 0$. Therefore,

$$
\begin{aligned}
\lim_{\tau \to \infty} \mathbf{Q}_x &= \lim_{\tau \to \infty} \mathbf{V}_x - \mathbf{V}_x\mathbf{A}'\mathbf{W}^{-1}\mathbf{A}\mathbf{V}_x + \mathbf{V}_x\mathbf{A}'\mathbf{W}^{-1}\mathbf{Q}_y\mathbf{W}^{-1}\mathbf{A}\mathbf{V}_x \\
&= \mathbf{V}_x.
\end{aligned}
$$

$\square$

Theorem 11.3. *Let us consider two hidden resolution processes with all parameters equal except σ_x^2 and σ_y^2. If the ratio between σ_x^2 and σ_y^2 is the same for the two processes, then they have the same autocorrelation function.*

Proof. Redefine $\tau = \kappa\sigma_x^2$, $\sigma_y^2 = \gamma\sigma_x^2$, $\boldsymbol{\Psi}_x = \mathbf{V}_x/\sigma_x^2$ and $\boldsymbol{\Psi}_y = \mathbf{Q}_y/\sigma_y^2$. Then

$$\mathbf{B} = \mathbf{V}_x\mathbf{A}'\mathbf{W}^{-1} = \sigma_x^2\boldsymbol{\Psi}_x\mathbf{A}'(\sigma_x^2\mathbf{A}\boldsymbol{\Psi}_x\mathbf{A}' + \kappa\sigma_x^2\mathbf{I})^{-1} = \boldsymbol{\Psi}_x\mathbf{A}'(\mathbf{A}\boldsymbol{\Psi}_x\mathbf{A}' + \kappa\mathbf{I})^{-1}$$

Thus, $\mathbf{B}$ does not depend on σ_x^2. Moreover:

$$\begin{aligned}\mathbf{Q}_x &= \sigma_x^2\boldsymbol{\Psi}_x - \mathbf{B}[\sigma_x^2(\mathbf{A}\boldsymbol{\Psi}_x\mathbf{A}' + \kappa\mathbf{I}) - \gamma\sigma_x^2\boldsymbol{\Psi}_y]\mathbf{B}' \\ &= \sigma_x^2\{\boldsymbol{\Psi}_x - \mathbf{B}[(\mathbf{A}\boldsymbol{\Psi}_x\mathbf{A}' + \kappa\mathbf{I}) - \gamma\boldsymbol{\Psi}_y]\mathbf{B}'\}.\end{aligned}$$

Hence, each element of $\mathbf{Q}_x$ is proportional to σ_x^2. Consequently, the correlation matrix of $\mathbf{x}_{1:n_x}$ does not depend on σ_x^2 and σ_y^2 in the new reparameterization. Therefore, the correlation matrix depends on σ_x^2 and σ_y^2 only through $\gamma = \sigma_y^2/\sigma_x^2$. □

Theorem 11.4 presents an identifiability problem: when γ, the ratio between σ_y^2 and σ_x^2, goes to infinity at a rate proportional to the square of κ, the ratio between τ and σ_x^2, the covariance matrix $\mathbf{Q}_x$ converges to a constant positive definite matrix.

Theorem 11.4. $\lim_{\kappa\to\infty, \gamma\to\infty, \gamma/\kappa^2=c} \mathbf{Q}_x = \mathbf{C}$, *where* $\det(\mathbf{C}) > 0$.

Proof. Let us use the same reparameterization used in the proof of Theorem 11.3. In order to simplify notation, in this proof "lim" means the limit as κ and γ tend to infinity with γ/κ^2 held fixed. First, let us consider the behavior of $\mathbf{B}(\mathbf{A}\boldsymbol{\Psi}_x\mathbf{A}' + \kappa\mathbf{I})\mathbf{B}'$:

$$\begin{aligned}\lim \mathbf{B}(\mathbf{A}\boldsymbol{\Psi}_x\mathbf{A}' + \kappa\mathbf{I})\mathbf{B}' &= \lim \boldsymbol{\Psi}_x\mathbf{A}'\kappa^{-1}\mathbf{I}(\mathbf{A}\boldsymbol{\Psi}_x\mathbf{A}' + \kappa\mathbf{I})\kappa^{-1}\mathbf{I}\mathbf{A}\boldsymbol{\Psi}_x \\ &= 0.\end{aligned}$$

Therefore,

$$\begin{aligned}\lim \mathbf{Q}_x &= \sigma_x^2\lim\{\boldsymbol{\Psi}_x - \mathbf{B}[(\mathbf{A}\boldsymbol{\Psi}_x\mathbf{A}' + \kappa\mathbf{I}) - \gamma\boldsymbol{\Psi}_y]\mathbf{B}'\} \\ &= \sigma_x^2\boldsymbol{\Psi}_x + \sigma_x^2\lim \boldsymbol{\Psi}_x\mathbf{A}'\kappa^{-1}\gamma\boldsymbol{\Psi}_y\kappa^{-1}\mathbf{A}\boldsymbol{\Psi}_x \\ &= \sigma_x^2\boldsymbol{\Psi}_x + c\sigma_x^2\boldsymbol{\Psi}_x\mathbf{A}'\boldsymbol{\Psi}_y\mathbf{A}\boldsymbol{\Psi}_x = \mathbf{C}.\end{aligned}$$

As $\boldsymbol{\Psi}_x$ and $\boldsymbol{\Psi}_y$ are positive definite, then $\det(\mathbf{C}) > 0$. □

In order to illustrate Theorem 11.4, Figure 11.1 shows as dotted lines the autocorrelation functions for $\phi_x = 0.5, 0.9, 0.99$, for $\lambda = 1, 5, 10, 20, 100, 1000$, 10000, $\sigma_y^2 = 5\lambda^2$, $\phi_y = 0.0$, $\sigma_x^2 = 1.0$, and $m = 12$. As λ and σ_y^2 increase with σ_y^2/λ^2 held constant, the autocorrelation function increases and converges to the limiting autocorrelation function represented by the solid line and depicted in Theorem 11.4.

Theorems 11.2 and 11.4 have important consequences that point to the right specification of the priors for the parameters of HRMs. In order to illustrate this point, Figure 11.2 shows different representations of the likelihood function for λ and σ_y^2, the other parameters kept constant at their posterior means, for the Fraser River application that will be presented in

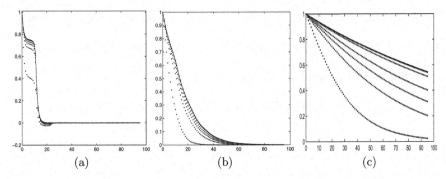

Fig. 11.1. Illustration of Theorem 11.4. Dotted lines are HRM autocorrelation functions with different values of λ $(1, 5, 10, 20, 100, 1000, 10000)$ and $\sigma_y^2 = 5\lambda^2$. As λ and σ_y^2 increase with σ_y^2/λ held constant, the autocorrelation function converges to the autocorrelation function represented by the solid line. In all figures, $\phi_y = 0.0$, $\sigma_x^2 = 1.0$, and $m = 12$. Panels (a), (b), and (c) correspond to different values of ϕ_x: (a) 0.5; (b) 0.9; (c) 0.99.

Section 11.6.1. The behavior of the likelihood is very interesting in two aspects. First, as can be seen in Figures 11.2 (a) and (b), when λ approaches infinity and σ_y^2 is kept constant, the likelihood does not vanish but becomes constant, equal to the likelihood of the original ARMA model $p(x)$. Second, as can be seen in Figures 11.2 (a) and (d), when λ and σ_y^2 approach infinity such that σ_y^2/λ^2 is constant, the likelihood function again does not vanish but approaches a constant value greater than zero. Hence, if the prior distribution for λ is improper, then its posterior distribution will also be improper. Nonetheless, as indicated in Figures 11.2(c) and 11.2(d), there is information in the likelihood function about the value of λ. Therefore, it is necessary to carefully assign a proper prior distribution for λ. In Section 11.3.2, a graphical study shows that the autocorrelation function is almost the same for $\lambda = 10$ and for $\lambda \to \infty$. Thus, assigning a proper prior is recommended for λ, with positive mass in the interval $(0, 10)$ and zero mass outside of the interval.

11.3.2 Autocorrelation Functions

This section analyzes plots of several possible autocorrelation functions to elucidate some interesting properties of Hidden Resolution Models. In addition, some of these properties are related to the analytical results obtained in Section 11.3.1 about the limiting behavior of the autocorrelation functions. As a by-product of the study of the autocorrelation functions, the parameters of Hidden Resolution Models are given suitable interpretations.

AR(1) Building Blocks

Let us now consider HRMs with an initial AR(1) process at the fine level and a revised AR(1) process at the coarse level. An interesting feature of

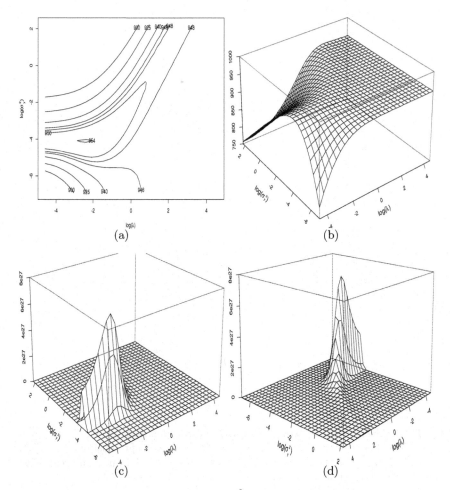

Fig. 11.2. Likelihood function for λ and σ_y^2: (a) log-likelihood contour plot; (b) log-likelihood; (c) likelihood; (d) likelihood from another perspective.

these models is the emulation of long memory processes (for references on long memory processes, see Beran, 1994, or Brockwell and Davis, 1991). As can be seen in Figure 11.3, the parameter ϕ_y controls the rate of decay of the autocorrelation function, which shows some persistence when the value of ϕ_y increases. When the number of hidden resolution levels is finite, HRMs do not have the long memory property but can emulate long memory processes. Nonetheless, their advantage over actual long memory models is interpretability; that is, the long memory type of behavior is explicitly modeled as a result of a high autocorrelation in the hidden coarse level of the hierarchy. Moreover, Figures 11.3 to 11.6 show that, when compared with traditional long memory models, Hidden Resolution Models provide richer autocorrelation structures.

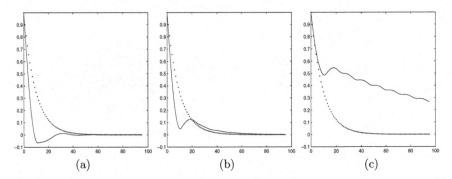

Fig. 11.3. AR(1) building blocks. Autocorrelation functions varying ϕ_y (the autoregressive coefficient on the coarse scale). Parameters kept constant: $\phi_x = 0.9$, $\sigma_x^2 = 1$, $\sigma_y^2 = 1$, $\lambda = 0.1$, and $m = 12$. Multiscale model (solid line); autoregressive model (dotted line). Particular values of ϕ_y: (a) 0.0; (b) 0.5; (c) 0.9.

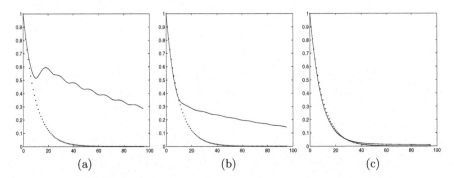

Fig. 11.4. AR(1) building blocks. Autocorrelation functions varying λ (the lack of agreement between coarse and fine levels). Parameters kept constant: $\phi_x = 0.9$, $\sigma_x^2 = 1$, $\phi_y = 0.9$, $\sigma_y^2 = 1$, and $m = 12$. Multiscale model (solid line); autoregressive model (dotted line). Particular values of λ: (a) 0.01; (b) 1.0; (c) 10.0.

   The values of λ control how much the generator mechanism of the coarse level influences the behavior of the fine level. Thus, as can be seen in Figure 11.4, the decay of the autocorrelation function also depends on λ. As shown in Section 11.3.1, the limit of the autocorrelation function when λ approaches infinity is the autocorrelation function of the original autoregressive process with coefficient ϕ_x. Actually, as can be seen in Figure 11.4, the autocorrelation functions of the HRM and original autoregressive process are already very close when $\lambda = 10$. When λ gets smaller, the model departs progressively further from the AR(1) model, as can be observed from the autocorrelation functions for $\lambda = 1$ and $\lambda = 0.01$.

   As shown in Section 11.3.1, the asymptotic behavior of the likelihood function when λ approaches infinity is a key feature of the model and has to be considered very carefully. When λ approaches infinity, the likelihood function

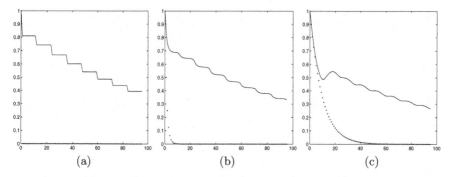

Fig. 11.5. AR(1) building blocks. Autocorrelation functions varying ϕ_x (the autoregressive coefficient on the fine scale). Parameters kept constant: $\sigma_x^2 = 1$, $\phi_y = 0.9$, $\sigma_y^2 = 1$, $\lambda = 0.1$, and $m = 12$. Multiscale model (solid line); autoregressive model (dotted line). Particular values of ϕ_x: (a) 0.0; (b) 0.5; (c) 0.9.

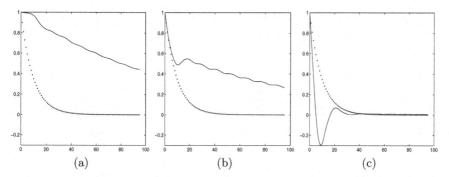

Fig. 11.6. AR(1) building blocks. Autocorrelation functions varying σ_x^2 (the variance of the error on the fine scale). Parameters kept constant: $\phi_x = 0.9$, $\phi_y = 0.9$, $\sigma_y^2 = 1$, $\lambda = 0.1$, and $m = 12$. Multiscale model (solid line); autoregressive model (dotted line). Particular values of σ_x^2: (a) 0.01; (b) 1; (c) 100.

approaches a constant equal to the likelihood function of an AR(1) process, implying that the use of an improper prior for λ would lead to an improper posterior. In addition, as shown in Figure 11.4, $\lambda = 10$ and $\lambda \rightarrow \infty$ provide almost the same autocorrelation function for the fine level. Therefore, a solution for the impropriety is to define a prior for λ truncated to the interval $(0, 10)$.

Another interesting feature of Hidden Resolution Models is the existence of a blocking effect that depends on m and ϕ_x. More specifically, when $\phi_x = 0.0$ the autocorrelations are constant by blocks of length m. Additionally, as can be seen in Figure 11.5, when ϕ_x increases the blocking effect is progressively reduced. Therefore, ϕ_x can be interpreted as the parameter that controls the smoothness of the autocorrelation function.

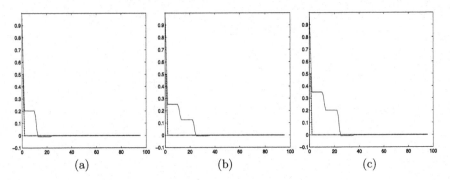

Fig. 11.7. MA(1) building blocks. Autocorrelation functions varying θ_y. Parameters kept constant: $\theta_x = 0.9$, $\sigma_x^2 = 1$, $\sigma_y^2 = 1$, $\lambda = 0.1$, and $m = 12$. Multiscale model (solid line); moving average model (dotted line). Particular values of θ_y: (a) 0.0; (b) 0.5; (c) 0.9.

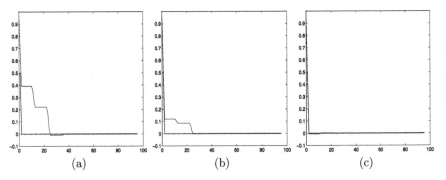

Fig. 11.8. MA(1) building blocks. Autocorrelation functions varying λ. Parameters kept constant: $\theta_x = 0.9$, $\sigma_x^2 = 1$, $\theta_y = 0.9$, $\sigma_y^2 = 1$, and $m = 12$. Multiscale model (solid line); moving average model (dotted line). Particular values of λ: (a) 0.01; (b) 1.0; (c) 10.0.

The decay of the autocorrelation function is also controlled by the parameters σ_x^2 and σ_y^2. As shown in Section 11.3.1, the autocorrelation function depends on σ_x^2 and σ_y^2 only through σ_x^2/σ_y^2. Therefore, it is necessary to study only the autocorrelation function for σ_x^2 or σ_y^2. Figure 11.6 depicts the autocorrelation function for σ_x^2 assuming values 0.01, 1, and 100. It is interesting to note that the decay of the autocorrelation function for the HRM can be faster than the decay for the original AR(1) process.

MA(1) Building Blocks

Let us now consider HRMs with an initial MA(1) process at the fine level and a revised MA(1) process at the coarse level. Some autocorrelation functions for these models are presented in Figures 11.7 to 11.10. For these models,

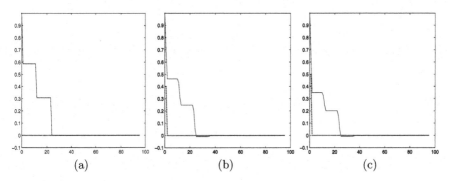

Fig. 11.9. MA(1) building blocks. Autocorrelation functions varying θ_x. Parameters kept constant: $\sigma_x^2 = 1$, $\theta_y = 0.9$, $\sigma_y^2 = 1$, $\lambda = 0.1$, and $m = 12$. Multiscale model (solid line); moving average model (dotted line). Particular values of θ_x: (a) 0; (b) 0.5; (c) 0.9.

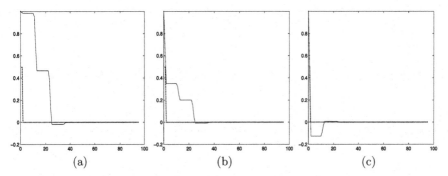

Fig. 11.10. MA(1) building blocks. Autocorrelation functions varying σ_x^2. Parameters kept constant: $\theta_x = 0.9$, $\theta_y = 0.9$, $\sigma_y^2 = 1$, $\lambda = 0.1$, and $m = 12$. Multiscale model (solid line); moving average model (dotted line). Particular values of σ_x^2: (a) 0.01; (b) 1; (c) 100.

the autocorrelation functions behave as step functions ($\theta_x = 0$) or variations of step functions ($\theta_x > 0$). Qualitatively, the autocorrelation functions for different parameter values are not very different from each other. Reflecting the MA(1) behavior at the coarse level, the autocorrelation functions vanish for lags greater than $3m$. From Figure 11.7, it can be seen that θ_y is related to the size of the autocorrelation function. As can be seen in Figure 11.8, λ can be interpreted as how close the HRM process is to the original MA(1) fine process. From Figure 11.9, θ_x is related to how fast the change in autocorrelation is for times closer to the block limits. Figure 11.10 shows that the size of the autocorrelation function also depends on the value of σ_x^2.

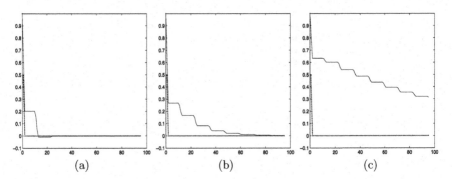

Fig. 11.11. AR(1) coarse level, MA(1) fine level. Autocorrelation functions varying ϕ_y. Parameters kept constant: $\theta_x = 0.9$, $\sigma_x^2 = 1$, $\sigma_y^2 = 1$, $\lambda = 0.1$, and $m = 12$. Multiscale model (solid line); moving average model (dotted line). Particular values of ϕ_y: (a) 0.0; (b) 0.5; (c) 0.9.

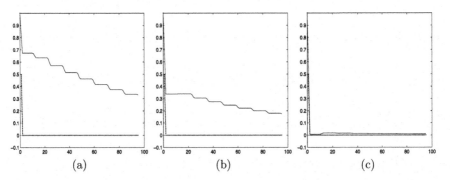

Fig. 11.12. AR(1) coarse level, MA(1) fine level. Autocorrelation functions varying λ. Parameters kept constant: $\theta_x = 0.9$, $\sigma_x^2 = 1$, $\phi_y = 0.9$, $\sigma_y^2 = 1$, and $m = 12$. Multiscale model (solid line); moving average model (dotted line). Particular values of λ: (a) 0.01; (b) 1.0; (c) 10.0.

AR(1) at the Coarse Level and MA(1) at the Fine Level

Let us now consider HRMs with an initial MA(1) process at the fine level and a revised AR(1) process at the coarse level. Some autocorrelation functions for these models are presented in Figures 11.11 to 11.14. As in the case of MA(1) building blocks, the autocorrelation functions behave as step functions ($\theta_x = 0$) or variations of step functions ($\theta_x > 0$). Moreover, the autocorrelation functions are similar for different parameter values. Reflecting the AR(1) process at the coarse level, the decay of the autocorrelation functions is slow for high values of ϕ_y. Interpretations of θ_x, σ_x^2, ϕ_y, and λ are the same as for the models with only AR(1) or only MA(1) building blocks.

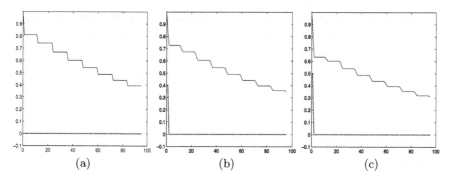

Fig. 11.13. AR(1) coarse level, MA(1) fine level. Autocorrelation functions varying θ_x. Parameters kept constant: $\sigma_x^2 = 1$, $\phi_y = 0.9$, $\sigma_y^2 = 1$, $\lambda = 0.1$, and $m = 12$. Multiscale model (solid line); moving average model (dotted line). Particular values of θ_x: (a) 0; (b) 0.5; (c) 0.9.

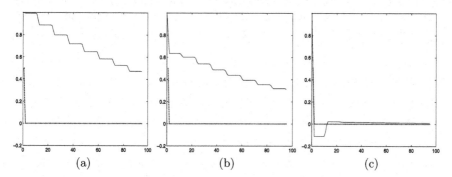

Fig. 11.14. AR(1) coarse level, MA(1) fine level. Autocorrelation functions varying σ_x^2. Parameters kept constant: $\theta_x = 0.9$, $\phi_y = 0.9$, $\sigma_y^2 = 1$, $\lambda = 0.1$, and $m = 12$. Multiscale model (solid line); moving average model (dotted line). Particular values of σ_x^2: (a) 0.01; (b) 1; (c) 100.

MA(1) at the Coarse Level and AR(1) at the Fine Level

Let us now consider HRMs with an initial AR(1) process at the fine level and a revised MA(1) process at the coarse level. Figures 11.15 to 11.18 depict autocorrelation functions for some of these models. This class of models provides several interesting autocorrelation functions. The autocorrelation functions are reasonably smooth for typical values of ϕ_x and die off for lags greater than $4m$. Moreover, they can oscillate smoothly between positive and negative values. For example, in Figure 11.18, we can see that when $\phi_x = 0.9$, $\theta_y = 0.9$, $\sigma_x^2 = 0.01$, $\sigma_y^2 = 1$, and $\lambda = 0.1$, the autocorrelation function initially falls slowly, then falls very quickly, becoming negative, and stabilizes at zero around lag $4m$. This autocorrelation function could represent, for example, a process that is being monitored to be kept at a given nominal level: it takes some time for people to realize the need for intervention; after the intervention, the process converges to the nominal level and slightly passes it; and finally (and hopefully) the process stabilizes at the nominal level.

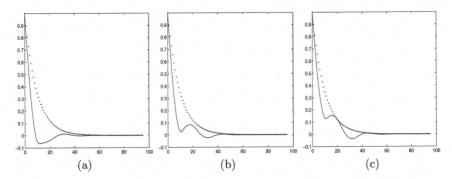

Fig. 11.15. MA(1) coarse level, AR(1) fine level. Autocorrelation functions varying θ_y. Parameters kept constant: $\phi_x = 0.9$, $\sigma_x^2 = 1$, $\sigma_y^2 = 1$, $\lambda = 0.1$, and $m = 12$. Multiscale model (solid line); autoregressive model (dotted line). Particular values of θ_y: (a) 0.0; (b) 0.5; (c) 0.9.

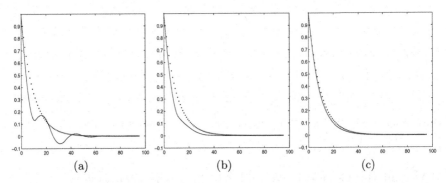

Fig. 11.16. MA(1) coarse level, AR(1) fine level. Autocorrelation functions varying λ. Parameters kept constant: $\phi_x = 0.9$, $\sigma_x^2 = 1$, $\theta_y = 0.9$, $\sigma_y^2 = 1$, and $m = 12$. Multiscale model (solid line); autoregressive model (dotted line). Particular values of λ: (a) 0.01; (b) 1.0; (c) 10.0.

11.4 Incorporating Periodicities

Periodicities are incorporated in MSTSMs and HRMs at the fine level by including regressors corresponding to harmonics with a cycle length equal to the coarsening window. As harmonics are sine and cosine functions with mean

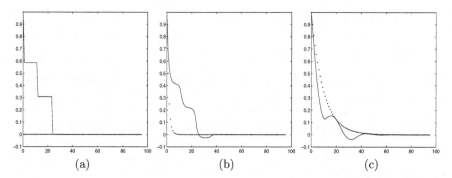

Fig. 11.17. MA(1) coarse level, AR(1) fine level. Autocorrelation functions varying ϕ_x. Parameters kept constant: $\sigma_x^2 = 1$, $\theta_y = 0.9$, $\sigma_y^2 = 1$, $\lambda = 0.1$, and $m = 12$. Multiscale model (solid line); autoregressive model (dotted line). Particular values of ϕ_x: (a) 0; (b) 0.5; (c) 0.9.

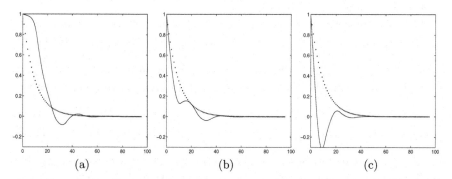

Fig. 11.18. MA(1) coarse level, AR(1) fine level. Autocorrelation functions varying σ_x^2. Parameters kept constant: $\phi_x = 0.9$, $\theta_y = 0.9$, $\sigma_y^2 = 1$, $\lambda = 0.1$, and $m = 12$. Multiscale model (solid line); autoregressive model (dotted line). Particular values of σ_x^2: (a) 0.01; (b) 1; (c) 100.

over the cycle length equal to zero, the coarsening operation eliminates the periodic pattern.

The construction of the model is the same as in Section 11.2.2, except that Equation (11.2) is substituted by $\mathbf{x}_{1:n_x} \sim N(\mathbf{x}_{1:n_x}|\boldsymbol{\mu}_x, \mathbf{V}_x)$ where $\boldsymbol{\mu}_x = \mathbf{Z}\boldsymbol{\beta}$ is a periodic vector with period equal to the coarsening window m, $\mathbf{V}_x$ is the covariance matrix of the fine-level process, and $\mathbf{Z}$ is a design matrix corresponding to the harmonics. It is easy to show that the updated marginal distribution of $\mathbf{x}_{1:n_x}$ is

$$q(\mathbf{x}_{1:n_x}) = N(\mathbf{x}_{1:n_x}|\mathbf{Z}\boldsymbol{\beta}, \mathbf{V}_x - \mathbf{B}(\mathbf{W} - \mathbf{Q}_y)\mathbf{B}'),$$

where $\mathbf{B} = \mathbf{V}_x\mathbf{A}'\mathbf{W}^{-1}$ and $\mathbf{W} = \mathbf{A}\mathbf{V}_x\mathbf{A}' + \mathbf{U}$ are the same as in Section 11.2.2.

Sections 11.6.1 and 11.6.3 present examples of the application of HRMs and MSTSMs to time series in the presence of periodicities.

11.5 Inference and Prediction

11.5.1 Posterior Simulation

The Bayesian estimation of the parameters of MSTSMs and HRMs cannot be performed analytically because the posterior distribution of the parameters is rather complicated due to nonlinearities. In order to explore the posterior distribution, an algorithm based on Markov chain Monte Carlo (MCMC) techniques is used to generate a sample from the posterior. This sample is then used to estimate summaries of the posterior distribution such as posterior means, standard deviations, and credible intervals.

A major aspect of the estimation for MSTSMs and HRMs is that, conditional on the hidden coarse level, the parameters corresponding to coarse and fine levels are independent. Thus, the inclusion of the generation of the hidden coarse level dramatically facilitates the implementation of a Gibbs sampler to explore the posterior distribution.

The following priors are assumed for the parameters: $\phi_y \sim N(m_{\phi_y}, S_{\phi_y})$, $\sigma_y^2 \sim IG(\nu_{\sigma_y}/2, \nu_{\sigma_y}s_{\sigma_y}/2)$, $\phi_x \sim TrN_{(-1.0,1.0)}(m_{\phi_x}, S_{\phi_x})$, $\sigma_x^2 \sim IG(\nu_{\sigma_x}/2, \nu_{\sigma_x}s_{\sigma_x}/2)$, $\lambda \sim TrIG_{(0,10)}(\nu_\lambda/2, \nu_\lambda s_\lambda/2)$, where $TrN_{(a,b)}$ and $TrIG_{(a,b)}$ denote respectively the normal and gamma distributions truncated to the interval (a, b). Then, the joint distribution of all involved random quantities is

$$p(\mathbf{x}_{1:n_x}|\mathbf{y}_{1:n_y}, \phi_x, \sigma_x^2, \lambda)q(\mathbf{y}_{1:n_y}|\phi_y, \sigma_y^2)p(\phi_y)p(\sigma_y^2)p(\phi_x)p(\sigma_x^2)p(\lambda). \quad (11.13)$$

From the joint distribution given by Equation (11.13), it is easy to verify that conditional on the hidden coarse level, the generation of the parameters corresponding to different levels can be done separately. Thus, as the coarse-level process is generally simple, techniques already available in the literature can be used to generate the coarse-level parameters. For example, if the coarse level follows an ARMA process, then the coarse-level parameters can be generated with the procedure proposed by Chib and Greenberg (1994).

The generation of the parameters of the fine level and link equation is not so trivial because, in general, the full conditional distributions are not available in closed form for sampling. To overcome this problem, these parameters are simulated with Metropolis-Hastings proposals. The following theorem simplifies and accelerates the computations for the simulation of $(\phi_x, \sigma_x^2, \lambda)$:

Theorem 11.5.

$$p(\phi_x, \sigma_x^2, \lambda|\mathbf{x}_{1:n_x}, \mathbf{y}_{1:n_y}) \propto p(\phi_x, \sigma_x^2)p(\mathbf{x}_{1:n_x}|\phi_x, \sigma_x^2)p(\lambda)$$
$$\times \frac{p(\mathbf{y}_{1:n_y}|\mathbf{x}_{1:n_x}, \phi_x, \sigma_x^2, \lambda)}{p(\mathbf{y}_{1:n_y}|\phi_x, \sigma_x^2, \lambda)},$$

where

$$p(\mathbf{y}_{1:n_y}|\mathbf{x}_{1:n_x}, \phi_x, \sigma_x^2, \lambda) = \prod_{s=1}^{n_y} N\left(y_s \middle| m^{-1}\sum_{i=1}^m x_{(s-1)m+i}, \lambda(\mathbf{A}'\mathbf{V}_x\mathbf{A})_{11}\right)$$

and

$$p(\mathbf{y}_{1:n_y}|\phi_x, \sigma_x^2, \lambda) = N(\mathbf{y}_{1:n_y}|\mathbf{0}, \mathbf{A}'\mathbf{V}_x\mathbf{A} + \mathbf{U}).$$

Proof.

Since the conditional distribution of $\mathbf{x}_{1:n_x}$ given $(\mathbf{y}_{1:n_y}, \phi_x, \sigma_x^2, \lambda)$ is not revised by Jeffrey's rule, then

$$q(\mathbf{x}_{1:n_x}|\mathbf{y}_{1:n_y}, \phi_x, \sigma_x^2, \lambda) = p(\mathbf{x}_{1:n_x}|\mathbf{y}_{1:n_y}, \phi_x, \sigma_x^2, \lambda)$$

$$= \frac{p(\mathbf{x}_{1:n_x}|\phi_x, \sigma_x^2, \lambda)p(\mathbf{y}_{1:n_y}|\mathbf{x}_{1:n_x}, \phi_x, \sigma_x^2, \lambda)}{p(\mathbf{y}_{1:n_y}|\phi_x, \sigma_x^2, \lambda)}.$$

The result follows on application of Bayes' Theorem. □

Note that $p(\mathbf{x}_{1:n_x}|\phi_x, \sigma_x^2, \lambda)$ and $p(\mathbf{y}_{1:n_y}|\mathbf{x}_{1:n_x}, \phi_x, \sigma_x^2, \lambda)$ are easy to compute and that the computation of $p(\mathbf{y}_{1:n_y}|\phi_x, \sigma_x^2, \lambda)$ can be done very efficiently by the Kalman filter (for a reference on the Kalman filter, see West and Harrison, 1997).

The generation of proposals for ϕ_x, σ_x^2, and λ is performed through Metropolis-Hastings steps. After the generation of each proposal, Theorem 11.5 is used to compute its acceptance probability. More specifically, the proposal for σ_x^2 is generated from $U(\sigma_x^2|\sigma_x^{2\,(old)}/\delta_{\sigma_x}, \sigma_x^{2\,(old)}\delta_{\sigma_x})$, where δ_{σ_x} has to be tuned to yield a reasonable acceptance rate.

The proposal for ϕ_x is simulated from $U\left(\phi_x|\max(-1.0, \phi_x^{(old)} - \delta_{\phi_x}), \min(10.0, \phi_x^{(old)} + \delta_{\phi_x})\right)$, where δ_{ϕ_x} has to be tuned to yield a reasonable acceptance rate.

In the case of MSTSMs, the generation of λ is easily performed. Conversely, in the case of HRMs, the hidden coarse level $\mathbf{y}_{1:n_y}$ and the parameter λ are highly correlated a posteriori. Thus, $\mathbf{y}_{1:n_y}$ and λ are jointly simulated in order to improve the mixing of the Gibbs sampler. First, a proposal $\lambda^{(new)}$ for λ is generated from $U(\lambda|\max(0, \lambda^{(old)} - \delta_\lambda), \min(1, \lambda^{(old)} + \delta_\lambda))$. Then, a proposal for $\mathbf{y}_{1:n_y}$ is simulated from its full conditional on $\lambda^{(new)}$. The joint proposal is accepted or rejected with the appropriate Metropolis-Hastings acceptance probability.

11.5.2 Forecasting

In order to compute forecasts, it is sufficient to obtain a sample from the predictive distribution of the future observations at the different levels. Point and interval forecasts can then be derived from this sample.

A two-stage procedure is used to generate each realization of the sample from the predictive distribution. First, a future realization of the hidden coarse level is simulated conditional on the past. After that, a realization of the fine level is simulated conditional on the past and on the realization of the coarse level. The following two theorems show that it is possible to sample from the coarse- and fine-level predictive distributions in a very efficient way.

The following theorem is very useful for the generation of predictions of the coarse level. The theorem states that the one-step-ahead predictive distribution at the coarse level depends on the whole past at the coarse level but depends only on the last observation at the fine level.

Theorem 11.6. $q(y_{n_y}|\mathbf{y}_{1:(n_y-1)}, \mathbf{x}_{1:(n_x-m)}) = q(y_{n_y}|\mathbf{y}_{1:(n_y-1)}, \mathbf{x}_{n_x-m}).$

Proof.

$$q(y_s|\mathbf{y}_{1:s-1}, \mathbf{x}_{1:[m(s-1)]}) \propto \int q(\mathbf{y}_{1:s})p(\mathbf{x}_{1:ms}|\mathbf{y}_{1:s})d\mathbf{x}_{(ms-m+1):ms}$$

$$\propto q(y_s|y_{s-1}) \int \frac{p(\mathbf{x}_{1:ms})p(\mathbf{y}_{1:s}|\mathbf{x}_{1:ms})}{p(\mathbf{y}_{1:s})}d\mathbf{x}_{(ms-m+1):ms}$$

$$\propto \frac{q(y_s|y_{s-1})}{p(\mathbf{y}_{1:s})} \int p(\mathbf{x}_{(ms-m+1):ms}|x_{m(s-1)})$$
$$\times p(y_s|\mathbf{x}_{(ms-m+1):ms})d\mathbf{x}_{(ms-m+1):ms}$$

$$\propto \frac{q(y_s|y_{s-1})}{p(y_s|\mathbf{y}_{1:(s-1)})}p(y_s|x_{m(s-1)}).$$

$\square$

In practice, the dependence on the past coarse level falls reasonably quickly with time.

The following theorem is important because it simplifies the task of forecasting the fine level when the future coarse level is known:

Theorem 11.7.

$$p(\mathbf{x}_{(n_x-m+1):n_x}|\mathbf{y}_{1:n_y}, \mathbf{x}_{1:(n_x-m)}) = p(\mathbf{x}_{(n_x-m+1):n_x}|y_{n_y}, x_{n_x-m}).$$

Proof.
Using the fact that $q(\mathbf{x}_{1:n_x}|\mathbf{y}_{1:n_y}) = p(\mathbf{x}_{1:n_x}|\mathbf{y}_{1:n_y})$, the conditional distribution of the fine level given the coarse level is not revised by Jeffrey's rule:

$$q(\mathbf{x}_{1:n_x}|\mathbf{y}_{1:n_y}) = p(\mathbf{x}_{1:n_x}|\mathbf{y}_{1:n_y}) \propto p(\mathbf{x}_{1:n_x})p(\mathbf{y}_{1:n_y}|\mathbf{x}_{1:n_x})$$

$$\propto p(x_1)\left[\prod_{i=2}^{n_x} p(x_i|x_{i-1})\right]\left[\prod_{j=1}^{n_y} p(y_j|\mathbf{x}_{(mj-m+1):(mj)})\right].$$

Therefore

$$p(\mathbf{x}_{(n_x-m+1):n_x}|\mathbf{y}_{1:n_y}, \mathbf{x}_{1:(n_x-m)}) \propto p(\mathbf{x}_{1:n_x}|\mathbf{y}_{1:n_y})$$

$$\propto p(y_{n_y}|\mathbf{x}_{(n_x-m+1):n_x}) \prod_{i=n_x-m+1}^{n_x} p(x_i|x_{i-1})$$

$$\propto p(\mathbf{x}_{(n_x-m+1):n_x}|y_{n_y}, x_{n_x-m}).$$

$\square$

Then, the following theorem follows from Theorem 11.7 by mathematical induction.

Theorem 11.8.

$$p(\mathbf{x}_{(n_x+1):(n_x+ml)}|\mathbf{y}_{1:(n_y+l)}, \mathbf{x}_{1:n_x}) = p(\mathbf{x}_{(n_x+1):(n_x+ml)}|\mathbf{y}_{(n_y+1):(n_y+l)}, x_{n_x}).$$

That is, conditional on the last observation x_{n_x} at the fine level and on the future observations $\mathbf{y}_{(n_y+1):(n_y+l)}$ at the coarse level, the future observations at the fine level are independent of the observations at the fine level up to time $n_x - 1$ and the observations at the coarse level up to time n_y.

11.5.3 Estimation and Forecasting in the Presence of Periodicities

The procedures for estimation and forecasting when there are periodicities in the fine level of the model are analogous to the case without periodicities.

The full conditional for β is $N(\mathbf{m}_\beta^*, \mathbf{C}_\beta^*)$, where $\mathbf{C}_\beta^* = (\mathbf{C}_\beta^{-1}+\mathbf{Z}'\mathbf{V}_x^{-1}\mathbf{Z})^{-1}$ and $\mathbf{m}_\beta^* = \mathbf{C}_\beta^*(\mathbf{C}_\beta^{-1}\mathbf{m}_\beta + \mathbf{Z}'\mathbf{V}_x^{-1}\mathbf{x}_{1:n_x})$.

The full conditionals for ϕ_y, σ_y^2, and τ are the same as in the case without periodicities. The full conditionals for ϕ_x and σ_x^2 are analogous to the case without periodicities but substituting $\mathbf{x}_{1:n_x}$ by $\mathbf{x}_{1:n_x}^* = \mathbf{x}_{1:n_x} - \mathbf{Z}\beta$.

The forecasting procedure in the case with periodicities is analogous to the case without them. More specifically, predictions are made for the series $\mathbf{x}_{(n_x+1):(n_x+l)}^*$, and the periodicities are added to these predictions.

11.5.4 Efficient Computation Using the Kalman Filter

Although $p(\mathbf{y}_{1:n_y})$ is substituted by $q(\mathbf{y}_{1:n_y})$ in the construction of the multiscale model, it is necessary to compute $p(\mathbf{y}_{1:n_y})$ in order to perform estimation and forecasting. This section discusses how to compute $p(\mathbf{y}_{1:n_y})$ and $p(y_s|\mathbf{y}_{1:(s-1)})$ very efficiently through the use of the Kalman filter.

The main point of this section is that the model $p(\mathbf{y}_{1:n_y})$ can be rewritten as a Dynamic Linear Model (DLM) (West and Harrison, 1997) and the Kalman filter can be used in order to compute $p(\mathbf{y}_{1:n_y})$ and $p(y_s|\mathbf{y}_{1:(s-1)})$. As our interest is the computation of the marginal $p(\mathbf{y}_{1:n_y})$, in this section it is assumed that the fine level is the unobservable state parameter of a DLM. First, it is necessary to rewrite Equations (11.1) and (11.3) as a DLM:

$$\mathbf{x}_s = \mathbf{G}\mathbf{x}_{s-1} + \mathbf{w}_s, \quad \mathbf{w}_s \sim N(\mathbf{0}, \mathbf{W}), \tag{11.14}$$

$$y_s = \mathbf{F}\mathbf{x}_s + v_s, \quad v_s \sim N(0, V), \tag{11.15}$$

where $\mathbf{x}'_s = (x_{(s-1)m+1}, \ldots, x_{sm})$, $\{\mathbf{W}\}_{ij} = \sigma_x^2 \phi_x^{|i-j|}(1 - \phi_x^{2\min(i,j)})/(1 - \phi_x^2)$, $\mathbf{F} = m^{-1}(1, \ldots, 1)$, $V = \tau\sigma_x^2$, and

$$\mathbf{G} = \begin{pmatrix} 0 & \cdots & 0 & \phi_x \\ 0 & \cdots & 0 & \phi_x^2 \\ \vdots & & \vdots & \vdots \\ 0 & \cdots & 0 & \phi_x^m \end{pmatrix}.$$

Define $D_0 = \{x \text{ is stationary}\}$ and $D_s = D_{s-1} \cap \{y_s\}$, $s = 1, \ldots, n_y$. Therefore, $\mathbf{x}_1|D_0 \sim N(\mathbf{a}_1, \mathbf{R}_1)$ with $\mathbf{a}_1 = \mathbf{0}$ and $\{\mathbf{R}_1\}_{ij} = \sigma_x^2 \phi_x^{|i-j|}/(1 - \phi_x^2)$.

The Kalman filter can be used to compute $p(y_s|D_{s-1})$, $p(\mathbf{x}_s|D_s)$ and $p(\mathbf{x}_{s+1}|D_s)$, $s = 1, \ldots, n_y$:

- $y_s|D_{s-1} \sim N(f_s, Q_s)$, where $f_s = \mathbf{F}\mathbf{a}_s$ and $Q_s = \mathbf{F}\mathbf{R}_s\mathbf{F}' + V$;
- $\mathbf{x}_s|D_s \sim N(\mathbf{m}_s, \mathbf{C}_s)$, where $\mathbf{m}_s = \mathbf{a}_s + \mathbf{A}_s e_s$, $\mathbf{C}_s = \mathbf{R}_s - \mathbf{A}_s Q_s \mathbf{A}'_s$, $\mathbf{A}_s = \mathbf{R}_s\mathbf{F}'Q_s^{-1}$, and $e_s = y_s - f_s$;
- $\mathbf{x}_{s+1}|D_s \sim N(\mathbf{a}_{s+1}, \mathbf{R}_{s+1})$, where $\mathbf{a}_{s+1} = \mathbf{G}\mathbf{m}_s$ and $\mathbf{R}_{s+1} = \mathbf{G}\mathbf{C}_s\mathbf{G}' + \mathbf{W}$.

Note that $p(y_s|\mathbf{y}_{1:(s-1)}) = p(y_s|D_{s-1})$ and $p(\mathbf{y}_{1:n_y}) = \prod_{s=1}^{n_y} p(y_s|D_{s-1})$.

11.6 Examples

Here several applications illustrate the power and the flexibility of HRMs and MSTSMs when applied to the analysis of time series from a variety of areas. The HRM is applied to three different time series. In Section 11.6.1, the HRM with seasonality is used to analyze the Fraser River dataset introduced in Chapter 3. Section 11.6.2 presents an analysis of temperatures of the Northern Hemisphere. Additionally, in Section 11.6.3, the MSTSM is used in an analysis of the potential hydroelectric energy in the southeast region of Brazil.

11.6.1 Analysis of the Flow of the Fraser River

This section presents an analysis of the flow of the Fraser River at Hope, British Columbia, as an example of the application of the HRM with seasonality. As described in Chapter 3, a standard time series analysis for the monthly data using harmonics suggests long memory dependence behavior for the monthly data but a simple AR(1) process for the annual data. That provides the motivation to perform an analysis using the HRM with seasonality.

Figure 3.1(a) presents the plot of the log of the mean monthly flows of the Fraser River from January 1913 to December 1990, which shows obvious seasonality and strong dependence between years. Thus a simple ARMA process with seasonality for the monthly series seems to be inappropriate, as it would probably give poor medium-term forecasts. Instead, we model this dataset with the HRM introduced in Section 11.2, as was done in Ferreira et al. (2006). In order to check the performance of the HRM in this particular application, its predictive performance is compared with the performance of an AR model.

An exploratory analysis has shown that the seasonality can be well explained by the first, fourth, and fifth harmonics, which will be included in the model as discussed in Section 11.4.

Figure 3.1(b) shows the plot of the monthly residuals after extracting the overall mean and the seasonality. Figure 3.2 shows the autocorrelation and partial autocorrelation functions of the monthly residuals, suggesting a long memory type process. As discussed in Section 11.3, this long memory behavior of the series can be captured by the HRM.

Figure 3.3 shows the autocorrelation and partial autocorrelation functions of the annual series, strongly suggesting an AR(1) process for the annual level of aggregation. Accordingly, in the application of the multiscale framework to the flow of the Fraser River, it is natural to assign the annual level of aggregation as the coarse level.

Posterior summaries for the parameters appear in Table 11.1. The estimation procedure proposed in Section 11.5 was used with a total of 5000 iterations of the Gibbs sampler. The tuning parameters of the Metropolis-Hastings proposals were set to provide reasonable acceptance probabilities, equal to 40%, 50%, and 46% for ϕ_x, τ, and λ, respectively.

Table 11.1. Fraser River Flow – Posterior summaries for the parameters.

	HRM		AR(1) model	
	Mean	Standard deviation	Mean	Standard deviation
ϕ_y	0.6562	0.1331	–	–
σ_y^2	0.0193	0.0075	–	–
ϕ_x	0.5958	0.0371	0.6093	0.0263
σ_x^2	0.0449	0.0023	0.0451	0.0021
τ	0.0106	0.0049	–	–
λ	0.5365	0.2369	–	–
β_1	−0.8422	0.0177	−0.8423	0.0179
β_2	−0.4612	0.0177	−0.4616	0.0176
β_3	0.3391	0.0116	0.3389	0.0116
β_4	−0.0565	0.0114	−0.0564	0.0114
β_5	−0.1014	0.0085	−0.1012	0.0086
β_6	0.0670	0.0086	0.0670	0.0085

A visual check of convergence was performed by running two MCMC chains with different starting values (for a discussion on convergence monitoring, see Gelman, 1996). Figure 11.19 shows the trace of ϕ_y, σ_y^2, ϕ_x, σ_x^2,

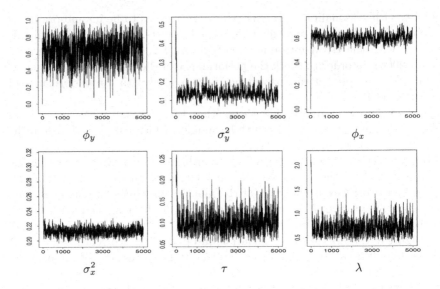

Fig. 11.19. Fraser River. Traces of ϕ_y, σ_y^2, ϕ_x, σ_x^2, τ, and λ.

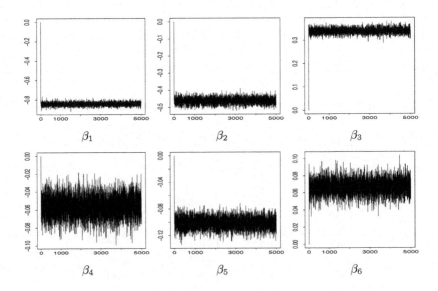

Fig. 11.20. Fraser River. Traces of the harmonic coefficients.

τ, and λ for a particular set of starting values. Figure 11.20 shows the trace of the harmonic coefficients. As it seems that the MCMC scheme converged after 1000 iterations, those first 1000 iterations were taken to be the burn-in period; the remaining 4000 iterations were used to perform the statistical

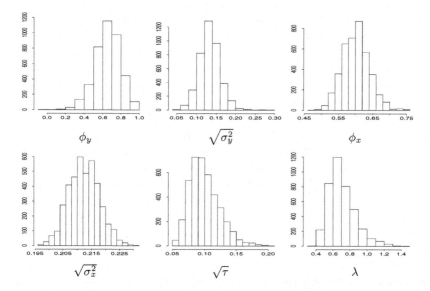

Fig. 11.21. Fraser River. Histograms of ϕ_y, $\sqrt{\sigma_y^2}$, ϕ_x, $\sqrt{\sigma_x^2}$, $\sqrt{\tau}$, and λ.

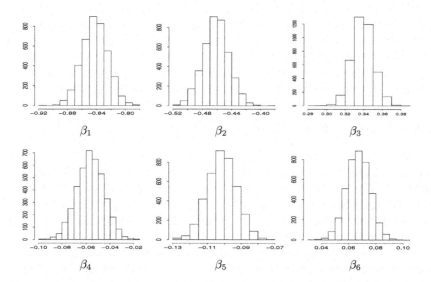

Fig. 11.22. Fraser River. Histograms of the harmonic coefficients.

inference presented here. Figures 11.21 and 11.22 show the histograms of the parameters of the model. As all the parameters are reasonably distant from zero, it is reasonable to keep all of them in the model.

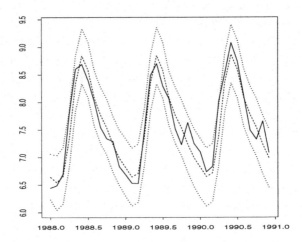

Fig. 11.23. Fraser River. Observed flow (solid), forecast (dashed), and predictive interval (dotted).

Figure 11.23 shows the observed monthly flow, the forecasts, and 95% predictive intervals for the years 1988, 1989, and 1990 using only the observations until 1987. As can be seen in Figure 11.23, the model performs very well in terms of predictive capacity. When compared with the AR(1) model, the HRM reduces the mean squared prediction error from 0.0549 to 0.0469, demonstrating for this example the superiority of the Hidden Resolution Model.

11.6.2 Northern Hemisphere Temperature

The analysis of the series of annual land-only average temperatures of the Northern Hemisphere is presented here as an example of an application of the HRM. Figure 11.24 shows the plot of annual land-only average temperatures of the Northern Hemisphere from 1851 to 1986 from the report of the Intergovernmental Panel on Climate Change (Houghton et al., 1990). Similar datasets were analyzed by Smith (1993) and Petris and West (1998) using long memory dependence models: Smith (1993) analyzed the series of monthly overall average temperatures of the Northern Hemisphere, and Petris and West (1998) analyzed the monthly overall average temperatures of the Southern Hemisphere. Here the HRM is used in order to emulate long memory dependence, assuming that the fine level is the annual time scale, and that there is an *unobservable* coarse level with coarsening window $m = 4$.

The MCMC algorithm described in Section 11.5.1 was used in order to fit the HRM. Figure 11.25 shows the posterior histograms of the parameters of the HRM. The posterior distribution of ϕ_y is very skewed, mainly because of

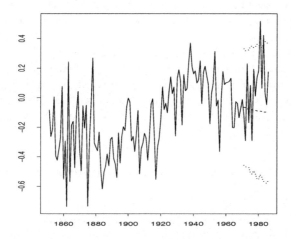

Fig. 11.24. Northern Hemisphere land temperatures. Observed series (solid), forecasts (dashed), and predictive intervals (dotted) for the period from 1971 to 1986.

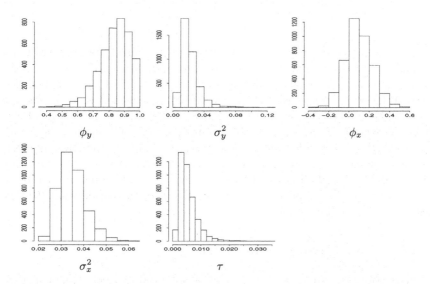

Fig. 11.25. Northern Hemisphere land temperatures. Posterior histograms of ϕ_y, σ_y^2, ϕ_x, σ_x^2, and τ.

the restrictions $-1 < \phi_y < 1$. The posterior distributions of σ_y^2, ϕ_x, σ_x^2 and τ are well behaved.

Table 11.2 presents posterior means and standard deviations for the parameters of both the HRM and the AR(1) models. There is a big difference between the posterior means for ϕ_x in the two models. This is expected when the HRM is significantly different from the AR(1) model such that ϕ_x has different interpretations in each of the models. The impact of the HRM can also be noticed by the magnitude of the posterior mean of ϕ_y, which is reasonably close to 1 as a result of the long memory type of behavior of the series. In addition, the posterior mean for τ is very small, imposing a high degree of agreement between the coarse and fine levels. Therefore, the introduction of an underlying coarse level seems to have a large impact on the process at the fine level.

Table 11.2. Northern Hemisphere temperatures – Posterior summaries for the parameters.

	HRM		AR(1) model	
	Mean	Standard deviation	Mean	Standard deviation
ϕ_y	0.8407	0.0955	–	–
σ_y^2	0.0235	0.0164	–	–
ϕ_x	0.1014	0.1284	0.4843	0.0816
σ_x^2	0.0352	0.0058	0.0465	0.0063
τ	0.0064	0.0042	–	–

In order to check the predictive capacity of the model, its forecasts were compared with those of autoregressive models. Data from 1851 to 1970 were used to fit both models, and then the forecasts from 1971 to 1986 were computed. When compared with the original AR(1) model, the HRM reduces the mean squared prediction error from 0.0936 to 0.0754, a reduction of about 19.4%. When compared with the best autoregressive model in this application, the AR(5), the reduction is about 2.8%. In addition, the HRM is more parsimonious, it has one parameter less than the AR(5) model, and therefore it is preferable. Finally, Figure 11.24 shows the observed annual average temperature, the predictive mean based on the HRM for the period from 1971 to 1986, and 95% predictive intervals.

11.6.3 Potential Hydroelectric Energy

In this section, the analysis of the series of potential hydroelectric energy of the rivers of the southeast region of Brazil is presented as an example of the application of the MSTSM with seasonality.

Figure 11.26(a) presents the plot of the monthly potential hydroelectric energy of the southeast region of Brazil from January 1937 to December 2000, as well as the series of annual averages. In the monthly series, the presence of seasonality and the strong dependence between the years are quite obvious. Moreover, the annual potential hydroelectric energy is well defined as the sum of the corresponding monthly series. Thus, the potential hydroelectric energy

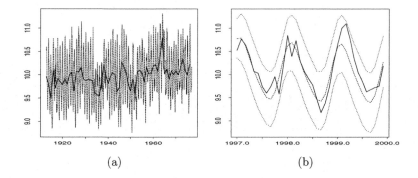

Fig. 11.26. Potential hydroelectric energy. (a) Monthly potential (dotted) and annual averages (solid). (b) Observed potential hydroelectric energy (solid), forecast (dashed), and predictive interval (dotted).

is modeled with the MSTSM with $\tau = 0$ and the annual series being the observed coarse level.

The model was compared with an AR(1) model with respect to its predictive capacity. Both models were fitted with the data until 1997 and used to compute three-years-ahead forecasts. The MSTSM performed much better than the AR(1) model in terms of mean squared prediction error, with values of 0.0532 and 0.0621 for the MSTSM and the AR(1) model, respectively, a decrease of 14.3% for the MSTSM. Furthermore, the mean squared prediction error is likely to decrease if there exists a model for the annual series with better predictive capacity. For example, a model for the coarse level can be built based on the knowledge of the physical process of the flow of the rivers in the southeast region of Brazil. Just as an indication of the potential increase in the predictive capacity, the mean squared prediction error would decrease by around 45.4% to 0.0339 if the future coarse level were known.

12

Change of Support Models

In contrast to the grid-based methods of the previous chapters in this section, the general class of methods often referred to as "change of support" methods evolved mostly to deal with problems of data that are observed at a scale different from the scale at which inference is desired, and the mapping between scales is nonregular. For example, demographic data may be available by postal code, but information about particular city blocks might be of interest. Alternatively, one could be trying to relate such demographic data to voting data, which are reported by voting precinct, and these precincts may not have any relationship to postal codes (so that precincts may overlap multiple postal codes, and postal code regions may contain multiple voting precincts). How is one to use the available data to draw valid conclusions at a different scale? The desire to answer this question is the primary motivation for the change of support problem.

Additional motivation comes from the desire to combine information at multiple scales. One typical example is attempting to combine individual medical record information with pollution data that are measured on a much larger scale. The methods must also deal with the case where the scales are nonnested, such as the postal codes and voting precinct example above.

The choice of scale and the method for changing scales can have major consequences for the results of the analysis. One of the earliest examples of the importance of the choice of scale was given by Gehlke and Biehl (1934), who were examining the relationship between male juvenile delinquency and median monthly rental rates by census tract in Cleveland. They discovered that the correlation coefficient between these two variables depended on whether the correlation was computed using the values from individual census tracts or using aggregated groups of tracts. In particular, if adjacent tracts were combined to form local clusters, the computed correlation coefficient was larger than when computed on the individual tracts. However, combining tracts in random groups did not impact the computed correlation. The authors also found similar results in the correlation between farm product value and the number of farmers in 1000 rural counties, with increasing correlation when counties were combined in contiguous blocks. A number of later authors have thoroughly investigated this phenomenon, and Gotway and Young (2002) provide a review of this literature.

Another vivid example was provided by Openshaw and Taylor (1979), who looked at the correlation between the percentages of Republican voters and elderly voters in Iowa. They started with data at the county level for the 99 counties in Iowa and then considered all possible aggregations into larger groups of these counties and computed the correlation using each of these possible datasets. They found that, using 12 groups, they could produce a correlation coefficient ranging from -0.97 to 0.99. Thus, with creative aggregation, the original data can be finagled to give just about any possible answer. Hence it is important to develop a methodology for aggregation (or for refining the scale) that is consistent with the true underlying process and that will accurately reflect that process at all scales.

A related concept from the epidemiology literature is that of "ecological inference", which describes learning about the behavior of individuals from aggregated data. It is analogous to trying to learn about a point-level process from regional data. However, as with the examples above, grouped data may not accurately reflect the relationship between variables at the individual level. Incorrectly drawing such an inference is referred to as the "ecological fallacy". Wakefield (2004) provides a recent discussion of this problem, which has by now generated extensive literature.

In the rest of this chapter, we present three methods for dealing with the change of support problem. The first assumes that there is an underlying process that occurs at the point level, and is typically modeled with Gaussian processes. The second is explicitly aimed at inference and prediction at an intermediate level, and also employs Gaussian processes. The third assumes that data occur inherently at the block level, with Markov random fields being a natural model for spatial processes. The basics of these spatial models were introduced in Chapter 2 and additional references can be found there. Gotway and Young (2002) is an extensive review article that provides a more general overview of the change of support literature (both Bayesian and non-Bayesian).

12.1 Point-Level Spatial Processes

A Gaussian process approach makes sense when the process can be viewed as arising from point data, with coarser-scale observations being block averages of the point process. For example, Banerjee et al. (2003) give an example on ground-level ozone pollution, which varies spatially and can be measured at individual points. One may then want to predict average pollution levels over an aggregated region, such as a postal code region, in order to relate it to other available data, such as hospital admittance data, that may only be obtainable at the postal code level. Thus, in order to study the relationship between hospital admittance and pollution, it is necessary to be able to estimate the average pollution over a region, and to do so from point measurements. The change of support methodology provides a coherent mechanism for moving between and among points and regions.

Some authors distinguish between various change of support problems, depending on the type of data that are observed, and what type of spatial inference is desired. A common distinction is to split the problem into four sub-problems: points-to-points, points-to-regions, regions-to-points, and regions-to-regions. In the first case, the process is observed at point locations, and inference is desired at other point locations. This is typically handled via ordinary kriging, as in Section 2.2. In the second case, inference on regions (blocks) is desired from point source data, and this is the primary focus of the development that follows. The methodology can then be extended similarly to deal with the other cases. The fourth case, regions-to-regions, is sometimes referred to as the modifiable areal unit problem.

In this section, we follow the approach of Banerjee et al. (2003, Ch. 6), and further details on these change of support methods can be found in that book and the references therein. The basic assumption is that there exists an underlying process $w(\mathbf{s})$ for $\mathbf{s} \in \mathcal{S}$, some region of interest, which will typically be a subset of Euclidean space. Point data would be observed or inferred over a set of locations $\{\mathbf{s}_1, \ldots, \mathbf{s}_m\}$. Block (region) data would be over a set of blocks $\{b_1, \ldots, b_n\}$. Because we assume that the basic process exists at the point level, block data are assumed to be derived as averages of the point-level process over the regions. Thus the block-level process is defined as

$$x(b_i) = \frac{1}{|b_i|} \int_{b_i} w(\mathbf{s}) d\mathbf{s}, \qquad (12.1)$$

where $b_i \subset D$ is a region of interest and $|b_i|$ is the area (or generalized volume) of the block b_i. This block averaging provides a statistically coherent method for moving between points and blocks. Possible alternatives would include taking the block value to be the process value at a central point or setting the block value to the average of the observed points in the block. However, such approaches tend to lead to larger variability and/or estimation bias when compared with Equation (12.1) (Banerjee et al., 2003).

To obtain the posterior predictive distribution for the block averages from point observations, we employ a Bayesian version of what is sometimes referred to as block kriging (Cressie, 1993; Chilès and Delfiner, 1999). Here we follow the Bayesian approach of Banerjee et al. (2003). For predicting over a set of blocks $\mathbf{b} = \{b_1, \ldots, b_n\}$ from data $\mathbf{y}$ at a set of points $\mathbf{s} = \{\mathbf{s}_1, \ldots, \mathbf{s}_m\}$, the predictive distribution is

$$f(\mathbf{x}(\mathbf{b})|\mathbf{y}(\mathbf{s})) = \int f(\mathbf{x}(\mathbf{b})|\mathbf{y}(\mathbf{s}), \boldsymbol{\theta}) f(\boldsymbol{\theta}|\mathbf{y}(\mathbf{s})) d\boldsymbol{\theta}, \qquad (12.2)$$

where $\boldsymbol{\theta}$ is the vector of parameters (both for the mean of the process and for the covariance structure). $f(\boldsymbol{\theta}|\mathbf{y})$ is the posterior distribution for the parameters, given the point-level observations, which would typically be fit using Markov chain Monte Carlo methods.

Denote the mean function of the Gaussian process by $\boldsymbol{\mu}$ and the covariance between points $\mathbf{s}_i$ and $\mathbf{s}_j$ by $\sigma^2 \rho(\mathbf{s}_i, \mathbf{s}_j)$. The Gaussian process model gives

$$f\left(\left(\begin{array}{c} \mathbf{y}(\mathbf{s}) \\ \mathbf{x}(\mathbf{b}) \end{array}\right) \Big| \boldsymbol{\theta}\right) = N\left(\left(\begin{array}{c} \boldsymbol{\mu}_y(\boldsymbol{\theta}) \\ \boldsymbol{\mu}_x(\boldsymbol{\theta}) \end{array}\right), \sigma^2 \left(\begin{array}{cc} \mathbf{H}_y(\boldsymbol{\theta}) & \mathbf{H}_{y,x}(\boldsymbol{\theta}) \\ \mathbf{H}_{y,x}^t(\boldsymbol{\theta}) & \mathbf{H}_x(\boldsymbol{\theta}) \end{array}\right)\right), \qquad (12.3)$$

where the jth element of the mean function vector for the blocks is

$$\mu_x(\boldsymbol{\theta})_j = E\left[x(b_j)|\boldsymbol{\theta}\right] = \frac{1}{|b_j|} \int_{b_j} w(\mathbf{s}|\boldsymbol{\theta}) d\mathbf{s},$$

the correlation matrix for the process observed at the vector of points $\mathbf{s}$ is $\mathbf{H}_y(\boldsymbol{\theta})$ with elements $\rho(\mathbf{s}_i, \mathbf{s}_j; \boldsymbol{\theta})$, the cross-correlation matrix between observed points and predicted blocks has elements (i, j) for the correlation between point $\mathbf{s}_i$ and block b_j,

$$\mathbf{H}_{y,x}(\boldsymbol{\theta})_{i,j} = \frac{1}{|b_j|} \int_{b_j} \rho(\mathbf{s}_i, \mathbf{t}; \boldsymbol{\theta}) d\mathbf{t},$$

and the correlation matrix for the block predictions has (i, j)th element

$$\mathbf{H}_x(\boldsymbol{\theta})_{i,j} = \frac{1}{|b_i| |b_j|} \int_{b_i} \int_{b_j} \rho(\mathbf{s}, \mathbf{t}; \boldsymbol{\theta}) d\mathbf{t} d\mathbf{s}.$$

Standard properties of the multivariate normal distribution imply that the conditional posterior predictive distribution is

$$\mathbf{x}(\mathbf{b})|\mathbf{y}(\mathbf{s}), \boldsymbol{\theta} \sim N\left(\boldsymbol{\mu}_x(\boldsymbol{\theta}) + \mathbf{H}_{y,x}^t(\boldsymbol{\theta})\mathbf{H}_y^{-1}(\boldsymbol{\theta})\left(\mathbf{y}(\mathbf{s}) - \boldsymbol{\mu}_y(\boldsymbol{\theta})\right), \qquad (12.4)\right.$$
$$\left. \sigma^2 \left[\mathbf{H}_x(\boldsymbol{\theta}) - \mathbf{H}_{y,x}^t(\boldsymbol{\theta})\mathbf{H}_y^{-1}(\boldsymbol{\theta})\mathbf{H}_{y,x}(\boldsymbol{\theta})\right]\right).$$

Unfortunately, these integrals typically cannot be analytically evaluated. Banerjee et al. (2003) suggest numerical integration via a Monte Carlo scheme. For each region b_j, they propose drawing a random (independent and uniformly distributed) sample of M_j points $\mathbf{s}_j^* = \{\mathbf{s}_{j,1}^*, \ldots, \mathbf{s}_{j,M_j}^*\}$ in b_j, with M_j possibly related to $|b_j|$ (so that larger regions may use more points for better accuracy). The numerical estimates necessary are then

$$\widehat{\mu}_x(\boldsymbol{\theta})_j = \frac{1}{M_j} \sum_{k=1}^{M_j} w(\mathbf{s}_{j,k}^*|\boldsymbol{\theta}),$$

$$\widehat{\mathbf{H}}_{y,x}(\boldsymbol{\theta})_{i,j} = \frac{1}{M_j} \sum_{k=1}^{M_j} \rho(\mathbf{s}_i, \mathbf{s}_{j,k}^*; \boldsymbol{\theta}),$$

$$\widehat{\mathbf{H}}_x(\boldsymbol{\theta})_{i,j} = \frac{1}{M_i M_j} \sum_{k=1}^{M_i} \sum_{l=1}^{M_j} \rho(\mathbf{s}_{ik}^*, \mathbf{s}_{jl}^*; \boldsymbol{\theta}).$$

Note that the accuracy of this estimation step depends on the size of the Monte Carlo sample, not the size of the original dataset. A single Monte Carlo sample can be generated for the entire set of blocks $\mathbf{b}$ and used for all of the estimates above.

Similarly, if one observed block data and wanted to predict at points, a version analogous to Equation (12.4) could be obtained from Equation (12.3) by focusing on $\mathbf{y}(\mathbf{s})$:

$$\mathbf{y}(\mathbf{s})|\mathbf{x}(\mathbf{b}), \boldsymbol{\theta} \sim N\left(\boldsymbol{\mu}_y(\boldsymbol{\theta}) + \mathbf{H}_{y,x}(\boldsymbol{\theta})\mathbf{H}_x^{-1}(\boldsymbol{\theta})\left(\mathbf{x}(\mathbf{b}) - \boldsymbol{\mu}_x(\boldsymbol{\theta})\right),\right.$$
$$\left.\sigma^2\left[\mathbf{H}_y(\boldsymbol{\theta}) - \mathbf{H}_{y,x}(\boldsymbol{\theta})\mathbf{H}_x^{-1}(\boldsymbol{\theta})\mathbf{H}_{y,x}^t(\boldsymbol{\theta})\right]\right).$$

Finally, to convert from one set of blocks to a different set of blocks (the modifiable areal unit problem), a point estimate for the value of a new block b_0 is the posterior mean of the process over that block,

$$E\left[x(b_0)|\mathbf{x}(\mathbf{b})\right] = E\left[\mu(b_0; \boldsymbol{\theta}) + \mathbf{H}_{x,b_0}^t(\boldsymbol{\theta})\mathbf{H}_x^{-1}(\boldsymbol{\theta})\left(\mathbf{x}(\mathbf{b}) - \boldsymbol{\mu}_x(\boldsymbol{\theta})\right)\Big| \mathbf{x}(\mathbf{b})\right],$$

where $\mathbf{H}_{x,b_0}(\boldsymbol{\theta})$ is the vector of correlations between $x(b_i)$ and $x(b_0)$ for $i \in \{1, \ldots, n\}$. A first-order approximation for the mean is given by

$$\mu(b_0; \boldsymbol{\theta}) \approx \frac{1}{|b_0|}\sum_{i=1}^{n}|b_i \cap b_0| \, x(b_i),$$

which is found by approximating the process as being roughly constant within each of the existing blocks b_i.

12.2 Inferring Intermediate-Level Processes

For some problems, specification of a complete probabilistic model at all possible scales may not be feasible, yet it is still necessary to draw a coherent inference from data at multiple scales. Wikle and Berliner (2005) provide methodology for such a situation. Suppose that data $\mathbf{y}_a = \{y(a_1), \ldots, y(a_{n_a})\}$ are observed on a fine set of blocks $\{a_1, \ldots, a_{n_a}\}$, and that coarse-scale data $\mathbf{y}_c = \{y(c_1), \ldots, y(c_{n_c})\}$ are observed on a set of blocks $\{c_1, \ldots, c_{n_c}\}$, with $n_c < n_a$. These data points are assumed to arise from noisy observations of two related underlying true processes, $\mathbf{x}_a$ and $\mathbf{x}_c$, i.e., $\mathbf{y}_a = \mathbf{x}_a + \boldsymbol{\varepsilon}_a$, where $\boldsymbol{\varepsilon}_a$ is mean zero with covariance matrix $\boldsymbol{\Sigma}_a$, and similarly $\mathbf{y}_c = \mathbf{x}_c + \boldsymbol{\varepsilon}_c$ with associated covariance matrix $\boldsymbol{\Sigma}_c$. Typically, $\boldsymbol{\Sigma}_a$ and $\boldsymbol{\Sigma}_c$ will be taken to be a constant times the identity matrix, i.e., errors are assumed to be independent and identically distributed, but the development here is sufficiently general to allow correlated error structures. This setup can be used for a variety of types of data, including those that arise from aggregation over blocks and those that are point measurements taken over different domains (e.g., predicted wind speeds from two different atmospheric models that give output on different regular lattices).

The goal here is inference or prediction at a potentially different scale, typically somewhere between the two other scales. By assuming that there are at most three scales of interest, the model can be built in a more restricted fashion without the need for the ability to model at an arbitrary scale. This will allow for various simplifications.

Denote the blocks on the level of interest as $\{b_1, \ldots, b_{n_b}\}$, with $n_c < n_b < n_a$. Thus the goal is to find the latent process $\mathbf{x}_b = \{x(b_1), \ldots, x(b_{n_b})\}$. This modeling is done in two steps, which separate the modeling of $\mathbf{x}_b$ itself from the modeling of the finer-scale residuals conditional on $\mathbf{x}_b$. Starting with the second of those, it is assumed that there exists a continuously defined process $x(\mathbf{s})$ for $\mathbf{s} \in D$ such that for each b_j and each $\mathbf{s} \in b_j$

$$x(\mathbf{s}) = x(b_j) + \gamma(\mathbf{s}),$$

where $\gamma(\mathbf{s})$ is a zero-mean process with a covariance function (defined over all of the space of interest, D) that does not depend on $\mathbf{x}_b$. Note that this formulation implies constant means within blocks, i.e., $E[x(\mathbf{s})] = E[x(b_j)]$.

The final piece of the methodology is the specification of the rest of the hierarchical model. A probability distribution, or at least the first two moments, must be specified for $\mathbf{x}_a$ and $\mathbf{x}_c$ given $\mathbf{x}_b$, showing how to relate the scales. The process $\mathbf{x}_b$ itself needs a specification. And then the parameters for each of these distributions, as well as for the $\mathbf{y}_a$ and $\mathbf{y}_c$ distributions, must be given appropriate hyperpriors. As full MCMC may be difficult or infeasible in large problems, Wikle and Berliner propose integrating out $\mathbf{x}_a$ and $\mathbf{x}_c$ when possible, reducing the dimension of the parameter space.

The relation of $\mathbf{x}_a$ and $\mathbf{x}_c$ to $\mathbf{x}_b$ is a key part of the specification. Areal averaging is a natural choice. In the simplest case when $a_i \subset b_j$, this implies that

$$x(a_i) = \frac{1}{|a_i|} \int_{a_i} x(b_j) d\mathbf{s} + \frac{1}{|a_i|} \int_{a_i} \gamma(\mathbf{s}) d\mathbf{s}$$
$$= x(b_j) + \frac{1}{|a_i|} \int_{a_i} \gamma(\mathbf{s}) d\mathbf{s}.$$

In the nonnested case, the first term is replaced by a weighted average,

$$x(a_i) = \frac{1}{|a_i|} \sum_{j=1}^{n_b} \int_{a_i \cap b_j} x(b_j) d\mathbf{s} + \frac{1}{|a_i|} \int_{a_i} \gamma(\mathbf{s}) d\mathbf{s}$$
$$= \left(\mathbf{g}_a^i\right)^t \mathbf{x}_b + \frac{1}{|a_i|} \int_{a_i} \gamma(\mathbf{s}) d\mathbf{s},$$

where $\mathbf{g}_a^i$ is the vector of weights with elements $|a_i \cap b_j|/|a_i|$, most of which will be zero. Similarly, the coarser process is obtained by

$$x(c_k) = \frac{1}{|c_k|} \sum_{j=1}^{n_b} \int_{c_k \cap b_j} x(b_j) d\mathbf{s} + \frac{1}{|c_k|} \int_{c_k} \gamma(\mathbf{s}) d\mathbf{s}$$

$$= \left(\mathbf{g}_c^k\right)^t \mathbf{x}_b + \frac{1}{|c_k|} \int_{c_k} \gamma(\mathbf{s}) d\mathbf{s}\,.$$

Additional details, as well as a full analysis of their motivating example in meteorology, can be found in Wikle and Berliner (2005). They also discuss a number of extensions of this methodology, including data observed at the intermediate level, spatio-temporal processes, multivariate processes, and non-additive error structures.

12.3 Block-Level Spatial Processes

In some cases, it does not make sense to think of a point-level process. Examples include data that are proportions (such as the proportion of voters favoring a candidate, which would be a binary response at the point level) or counts over regions. In these cases, it is necessary to operate solely at the block level, without recourse to an underlying point-level process. This situation is another example of the modifiable areal unit problem. Banerjee et al. (2003) distinguish between two types of these problems. The first case is when the new blocks (the *target* blocks) are contained within the set of existing blocks (the *source* blocks), and they call this "nested block realignment". The second case, nonnested block realignment, covers the more general situation where neither set of blocks necessarily contains the other set.

A simple approach to the nested realignment problem is to assume that the process is approximately constant within blocks, and then the new target block can be computed as a linear interpolation. For example, in the case of measuring counts (e.g., of people or events) in blocks and then denoting a new target block by b_0 and the source blocks by $\{b_1, \dots, b_n\}$, with $|b|$ denoting the area or generalized volume of a block, the value for b_0 can be approximated by

$$x(b_0) \approx \sum_{i=1}^{n} \frac{|b_0 \cap b_i|}{|b_i|} x(b_i)\,.$$

However, in many cases the assumption of constant rates within blocks is unrealistic. Furthermore, it is difficult to obtain an estimate of the uncertainty from this formulation. Flowerdew and Green (1989, 1994) developed models for incorporating covariates and using these covariates for improved estimation of the new target block values. Building on these models, Mugglin and Carlin (1998) developed a fully Bayesian approach that uses hierarchical models for predicting target block values. The exact form of the model will depend on the problem at hand. Best et al. (2000) proposed a related approach based on a marked point process model.

The nonnested block realignment problem is a bit more complicated because of the possible edge cases. Denote the observed values by $\{y(b_1), \dots, y(b_n)\}$ on a set of nonoverlapping blocks (which we refer to as a grid of blocks)

$\{b_1, \ldots, b_n\}$. Covariate information may be available on an alternate set of blocks $\{c_1, \ldots, c_m\}$. The key is to partition the space into disjoint subblocks that are entirely within both a single b_i and a single c_j, so that a finer set of regions is defined that is as common as possible to both original sets of blocks (edge cases may prevent a single common grid when $\bigcup_{i=1}^{n} b_i \neq \bigcup_{j=1}^{m} c_j$, for example, when the boundaries of the grid of zip code blocks covering a metropolitan area might not match the boundaries of a grid defined by the counties comprising that metropolitan region). Thus, divide each b_i into regions $b_i = \bigcup_{j=0}^{m} (b_i \cap c_j)$, where c_0 is the region (if any) outside the grid; i.e., $c_0 = S - \bigcup_{j=1}^{m} c_j$. Note that many of the $b_i \cap c_j$ may be empty sets, so that a more efficient numbering scheme would index over only the nonempty subsets. For notational convenience, denote these intersection sets by B_{ij}, which Banerjee et al. (2003) call *atoms*. Note that these atoms may be disconnected. Similar labeling is done using the covariate blocks as the reference; i.e., C_{kl} is the lth (nonempty) intersection of c_k with one of the b blocks. Thus, each nonedge atom will have two equivalent labels (a B_{ij} and a C_{kl}), whereas edge atoms will exist with only one label. Because there may be both B and C edge atoms, no simple single indexing scheme can be used.

With these indices, Banerjee et al., following Mugglin et al. (2000), describe a fully Bayesian model for inference and prediction, incorporating covariates measured on either or both grids, as well as a spatial process that is specified for the mean levels of the blocks (for both b and c). They suggest a Markov random field specification for these latent spatial process, and assume that these values are inheritable; i.e., the same value can be used for all of the relevant atoms of a B_{ij} or a C_{kl} cell. They also assume that the covariates are either aggregates (which can thus be apportioned) or inheritable measurements. They require that the $y(b_i)$'s be aggregated measurements (such as counts), so that these values can be apportioned among the B_{ij} atoms. In the case of counts, a Poisson formulation is used. For continuous responses, a gamma formulation could be used, leading to latent scaled product Dirichlet distributions on the atoms. Alternatively, a normal model could be used, with latent conditional multivariate normals. Banerjee et al. (2003) provide a detailed example of the Poisson case.

Part IV

Implicit Multiscale Models

13

Implicit Computationally Linked
Model Overview

In contrast to the rest of the multiscale models discussed in this book, the methods of this section do not attempt to explicitly model the joint distribution of the data at all of the scales. Instead, a probability model is specified for each scale independently, and then the scales are linked during the fitting algorithms. A fully dependent multiscale fit is then obtained implicitly.

This may seem like a strange approach to multiscale modeling. In fact, most of the methods of this section were originally developed to aid in the fitting of highly multimodal single-scale models. One way to attempt to deal with the multimodality is to move to a coarser resolution, which has an effect similar to smoothing the function and thus ameliorating the multimodality. In dealing with a likelihood, maximization is thus aided by improving the chances of finding all of the primary modes, including the global maximum. The speed of finding these modes can also be greatly increased. For a fully Bayesian posterior analysis, MCMC is typically employed. Standard Gibbs and Metropolis-Hastings steps can have great difficulty escaping from local modes, leaving exploration of the space incomplete. Entire modes can easily be missed, causing sections of the posterior density to be left out. By coarsening, the mixing of the chain can be greatly improved, allowing more complete exploration of the space in a smaller amount of computing time.

If multiple scales are used merely as a computational tool, then the ultimate object of interest is only the finest scale. In the case of optimization routines, it is only the end result that matters, when the algorithm has converged at the finest scale. For MCMC runs, if the joint multiscale space is explored, then the marginal distribution at the finest scale is easily obtained by discarding all information from the other scales. It was in this sort of context that most of these methods were created and refined.

On the other hand, as with the rest of this book, one may be truly interested in a multiscale problem, either with data observed at multiple scales, or with only unobservable yet physically meaningful processes at more than one scale. It turns out that most of the methods of this section are also directly applicable to these multiscale problems. They provide a computationally efficient approach to dealing with multiple scales in the presence of multi-modality, which is a frequent concern.

The rest of this chapter will discuss approaches leading up to multigrid methods and simulated sintering. The latter grew out of simulated annealing, a stochastic optimization technique, and in between is simulated tempering, a related optimization technique that uses varying amounts of smoothness to help deal with multimodality. Simulated sintering varies the scale to allow better exploration at a coarser scale and then refines the scale over time to hopefully eventually settle at the global maximum. Sintering is basically just a computational tool and is not as useful for a true multiscale problem. However, it is closely related to some of the other methods presented here, so we include it for perspective.

The second set of methods are multiscale MCMC algorithms that involve Metropolis-Hastings steps between scales, so that information moves directly between the scales, inducing posterior dependence. Standard Gibbs and Metropolis steps are mixed with these multiscale swapping steps to explore the joint multiscale posterior. These methods are equally applicable to multimodal single-scale problems and true multiscale problems.

The final methods are genetic algorithms and related MCMC algorithms. Genetic algorithms are a class of optimization techniques inspired by the action of genes on chromosomes. A population of solution chromosomes is evolved through the sharing of gene information between the chromosomes. A selection mechanism favors the better solutions. A series of iterations tends to converge to the global maximum if the initial population is selected from a wide enough set. This approach can also be adapted for MCMC by creating Metropolis proposals based on the mechanics of genetic algorithms. Such algorithms can be used for either optimization or full posterior inference for a truly multiscale problem.

Before moving on, we note that there are alternative computational approaches to using coarse-scale samples to help explore the fine-scale posterior. One basic alternative would be importance sampling (Evans and Swartz, 1995). Here, MCMC could be used to sample from the coarse posterior and to construct fine-scale candidates, which could then be used as the importance sample. However, on even simple one-dimensional examples, this approach can have serious problems. In particular, the importance weights tend to be dominated by one or two huge weights. This is because the importance sample does a poor job covering the region of highest posterior density on the fine scale. The few realizations from the importance distribution that do hit this region yield huge importance weights. Such problems can be expected since the coarse-to-fine proposal is not designed to use information from the likelihood.

Another alternative is the reversible jump approach of Green (1995). Here, MCMC draws are constructed from a mixture of the posteriors at each scale. Note that this new target distribution is a mixture of densities whose support may be on different spaces, adding to the complexity. We shall not focus on such methods here but instead refer the reader to additional references (Carlin and Chib, 1995; Richardson and Green, 1997; Brooks, 1998; Besag, 2000).

13.1 Simulated Annealing

Simulated annealing is a maximization technique developed for multimodal problems (Kirkpatrick et al., 1983). In such problems, the standard deterministic algorithms (such as gradient descent) are prone to becoming trapped in a local maximum. While these algorithms are typically optimal for well-behaved problems, they can fail spectacularly in the more complex problems often found in real applications. Thus some element of randomness is necessary during the maximization to facilitate escapes from local maxima in order to find the true global maximum.

Simulated annealing is one of the most popular of these approaches. The method draws analogies to statistical mechanics (equilibrium tends to occur in the minimum energy state) and the production of crystals and metals. In materials science, annealing refers to a process whereby a material is heated until it melts, and then slowly cooled to form a pure crystalline or metallic structure. Cooling must happen sufficiently slowly, or else the atoms or molecules will settle too quickly and not in their optimal places. But if done at the right pace, the random movements of the particles will tend to lead them to settle down into the optimal structure, forming the desired substance. Simulated annealing is similar, in that the typically high-dimensional parameter is initially allowed to move around quite randomly, like the particles at high temperature. Then, as the system is cooled, the parameter tends to make smaller movements and eventually settle into a maximum (or, to be completely analogous, one would work with the negative of the function to be maximized, and instead minimize this function, just as annealing minimizes the energy state of the system).

The algorithm is based on analogy with the Boltzmann probability distribution, $P(E) \propto \exp(-E/kt)$, where E is the energy function, k is Boltzmann's constant, and t is the temperature. In simulated annealing, E is the function to be minimized (and thus $-E$ is maximized). Applied to find the maximum of a posterior (or likelihood), the $-E$ can be taken as the posterior or just a key piece of the posterior. For example, in the common case of a Gaussian likelihood (with a flat or conjugate prior),

$$f(\theta|y) \propto \exp\left\{ -\frac{1}{2\sigma^2} \sum_{i=1}^{n} (y - g(\theta))^2 \right\},$$

one could then choose

$$E = \sum_{i=1}^{n} (y - g(\theta))^2,$$

where $g(\theta) = \theta$ in the simplest case, but could also be a tremendously complicated function (such as in the hydrology example of Chapter 16). In this way, there is only a single exponentiation, and $P(E)$ is essentially taken to be the posterior.

The temperature t is started at a relatively large (relative, in a sense, to E) value and then slowly decreased to "cool" the system. The algorithm is started with an initial value for the parameters and thus for E. A new value of parameters and thus a new energy E^* is proposed, typically as a random perturbation (as with a Metropolis step in Markov chain Monte Carlo). This new value is accepted with probability $\min(1, P(E^*)/P(E))$. Thus, if the new value is an improvement, i.e., $P(E^*) > P(E)$, then the new value is always accepted, but if the new value is worse then it is only accepted with a probability that depends on just how much worse it is. "Worse" is a relative term, as it also depends on the temperature t. Thus, when t is large, small changes in E are relatively unimportant and the algorithm will move easily. As t decreases, changes in E become more significant and the system eventually stabilizes at a mode, hopefully the global mode. The schedule for decreasing t may require tuning for good performance of the algorithm.

Because this algorithm is not deterministic, in multimodal situations it typically outperforms greedy algorithms (those that deterministically optimize at each step). While it tends to seek maxima and will often initially find only a local maximum, it also has a chance of escaping in order to explore the rest of the space and find the global maximum. The key is that the temperature scale must be chosen so that local maxima can be escaped when the temperature is at its highest. If the maxima are well separated, then a high initial temperature is necessary. Thus the optimal choice of the temperature schedule will depend on the function being maximized. But, if done correctly, simulated annealing has been found to work well in practice on highly multimodal problems. For example, it was one of the first algorithms to perform well on the traveling salesman problem (Kirkpatrick et al., 1983).

The original Metropolis version of MCMC (Metropolis et al., 1953) can be seen as a version of simulated annealing where the process does not cool. So while in simulated annealing the goal is to find a point estimate for the maximum and the process cools to reach this stable point, in MCMC the goal is to estimate a posterior density. A sample of points is used to empirically estimate the posterior, where this sample is taken at a fixed temperature using the same sampling procedure as in simulated annealing.

13.2 Simulated Tempering

Simulated tempering, proposed independently by Marinari and Parisi (1992) and Geyer and Thompson (1995), builds on the ideas of simulated annealing by changing the cooling analogy from that of the movement of the chain to that of the equilibrium distribution itself. A rough description of the process is that MCMC is run but the posterior itself is started as a heated version and cooled during the process. The idea is that heated versions of the posterior are smoothed out so that the local maxima are reduced, allowing the chain to more easily explore the whole space and not become stuck in a local maximum. The heating is done by taking the posterior to a fractional power, thus keeping some of the original shape while diminishing the peaks.

In mathematical terms, an auxiliary parameter is introduced which indexes the "temperature" of the system. The temperature is defined as a power for the posterior; i.e., let $j \in \{1, \ldots, m\}$ be the auxiliary parameter, let $1 = \beta_1 < \beta_2 < \ldots < \beta_m$, and denote the unnormalized posterior (the product of the likelihood and the prior) by $h(\boldsymbol{\theta})$. Then $h_j(\boldsymbol{\theta})$ is defined as $h(\boldsymbol{\theta})^{1/\beta_j}$. A prior $\pi(j)$ must be specified for the different temperature states. Then MCMC is performed over the joint space $(\boldsymbol{\theta}, j)$ for $f(\boldsymbol{\theta}, j) = h_j(\boldsymbol{\theta})\pi(j)$:

1. update $f(\boldsymbol{\theta}|j) = h_j(\boldsymbol{\theta})$ via standard Gibbs or Metropolis-Hastings steps,
2. update $f(j|\boldsymbol{\theta})$ using Metropolis-Hastings with proposal $j_{new} = j_{old} \pm 1$ with either equal or unequal move probabilities,
3. repeat steps 1 and 2 until convergence and desired number of samples are obtained.

In the second step, the acceptance probability is $\min(1, \alpha)$, where

$$\alpha = \frac{h_{j_{new}} \pi(j_{new}) q(j_{new}, j_{old})}{h_{j_{old}} \pi(j_{old}) q(j_{old}, j_{new})}$$

and $q(j_{old}, j_{new})$ is the probability of moving from j_{old} to $j_{new} = j_{old} \pm 1$. If the probabilities of moving up and down by 1 are equal, then the q terms will drop from the ratio.

This approach allows the Markov chain to explore not just the original posterior but also smoothed versions of the posterior, which can lessen the probability of becoming stuck in local maxima. For final inference where one only cares about the "cool" function, one conditions on $j = 1$ by simply discarding all samples with $j \neq 1$.

Figure 13.1 shows the process of heating an objective function. The top graph is the target function to explore (it was built by summing seven normal densities and one gamma density, and then squaring to produce the sharp modes). Note that if an MCMC run found itself in either the leftmost or especially the rightmost mode, it would be quite difficult for it to escape from that local maximum and find the true mode. The second graph is a partially "heated" version which was created by taking the square root of the original function. In particular, note how the separation between the two modes on the left is much diminished so that one can now imagine moving between those two modes somewhat easily. The third and fourth graphs are again successively taking square roots, so that the bottom graph is the top graph to the one-eighth power. In that final graph, the locations of the modes are still clear to the viewer, yet the function is overall becoming close to constant over the central region of interest, so that an MCMC run would wander across that whole region, visiting all three modes regularly. This is analogous to a substance that is raised in temperature until it is melted, so that atoms or molecules can freely move around. As the substance is cooled, the atoms gradually fall into place in an orderly fashion, and in Figure 13.1 as the process is "cooled", the idea is for an MCMC run to eventually settle into the primary global mode at 0.3.

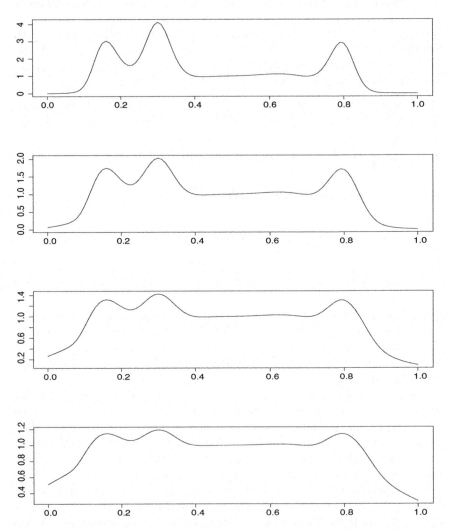

Fig. 13.1. The simulated tempering process. The top graph is the target function, and the lower graphs are successively heated versions which reduce the effect of the modes and allow better exploration of the function.

Geyer and Thompson (1995) suggest tuning the setup to obtain acceptance rates of moving between temperatures in the range of 20% to 40%. Such tuning can be done by adjusting the number of temperatures m, by changing the temperatures (the inverse powers β_j), and by wisely choosing the prior on the temperatures $\pi(j)$.

13.3 Simulated Sintering

Simulated tempering uses the concept of temperature to relate the target distribution to "heated" distributions that are smoother, so that an MCMC run is more easily able to traverse the space and find the global maximum. Simulated sintering (Liu and Sabatti, 1999) is analogous, except that instead of defining "hotter" as smoother, hotter is defined as a coarser scale. Thus we now can see how the whole simulated annealing framework relates to multiscale modeling. The target distribution is typically on a relatively fine scale or grid. Such a high-dimensional process can be difficult to effectively explore via MCMC, so successively "heated" versions use coarser and coarser grids. By coarsening, one can often reduce the effects of extrema, thus achieving a relatively flatter overall distribution (although in this case not necessarily "smoother", as the coarser grid tends to lead to sharper peaks and valleys than the more rounded graphs on finer grids). The term "sintering" is from metallurgy, where it relates to fractioning material into smaller portions.

A visual demonstration of sintering on the same 1-D function from Figure 13.1 appears in Figure 13.2, where the top graph is the same target function and the following graphs show successive coarsening. The coarsening reduces the differences between the modes and the rest of the graph. Thus increases in "temperature" correspond to coarser grids and thus decreased dimensionality of the target space.

Simulated sintering drew its inspiration from multigrid methods (see the next section) and from ideas in Wong (1995), who was one of the first in the statistical literature to propose an auxiliary variable for scale/coarseness as part of an MCMC run.

Mathematically, sintering is justified through and serves as just one example of group moves for generalized Gibbs sampling and MCMC. In standard Gibbs sampling, parameters are updated one at a time. The idea of a group move is to transform a collection of parameters by drawing a joint random move that leaves the equilibrium distribution invariant. For more details, the reader is referred to Liu and Sabatti (1999, 2000).

The execution of simulated sintering is analogous to that of simulated tempering. An auxiliary variable j is created to represent the current scale of the process, with its space $\mathcal{X}_j$ satisfying $\dim(\mathcal{X}_1) > \dim(\mathcal{X}_2) > \ldots > \dim(\mathcal{X}_m)$. For the space $\mathcal{X}_j$, the unnormalized posterior is given by $h_j(\boldsymbol{\theta}_j)$. Things can be kept simple by choosing a discrete uniform prior over the spaces. MCMC is then run over the joint space of $\boldsymbol{\theta}$ and j. Within a scale, updates

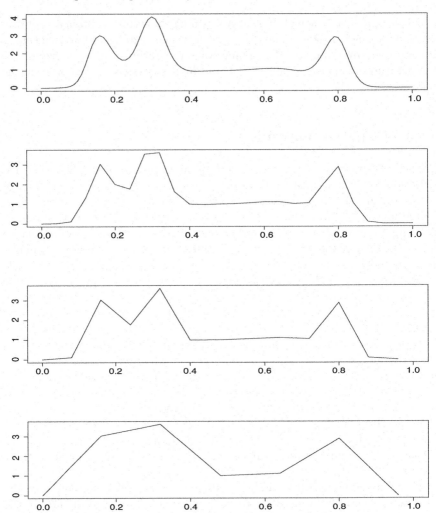

Fig. 13.2. The simulated sintering process. The top graph is the target function and lower graphs are successively heated versions which reduce the effect of the modes and allow better exploration of the function.

for $f(\boldsymbol{\theta}_j|j) = h_j(\boldsymbol{\theta}_j)$ are performed with standard moves. For moving between scales, reversible jump MCMC (Green, 1995) is used, which just includes an additional Hastings term (Hastings, 1970) that adjusts for the probabilities of proposals that move between spaces of different dimensions. Moves within and between scales are then alternated until convergence is achieved and the desired number of posterior samples is drawn.

13.4 Multigrid Methods

Multigrid methods (e.g., Goodman and Sokal, 1989; Briggs et al., 2000) were originally developed in the physics and applied mathematics literature, primarily for the solution of systems of differential equations. When such systems cannot be solved analytically (which is typical of complex systems), they must be solved numerically. This approach requires setting a grid on the parameter space over which the solution is found. After that, iterative methods based on single grid-block updates, such as Jacobi's algorithm, are used in order to find the solution. Defining the error as the difference between the correct solution and the solution at the present iteration, these algorithms are able to eliminate the high-frequency components of the error. Unfortunately, the low-frequency components of the error are not eliminated by these single grid-block update algorithms. Multigrid methods solve this problem by recognizing that low-frequency components at one resolution level become high-frequency components at coarser resolution levels. Moreover, coarser grids lead to inaccuracies in the solution, while fine grids require more computing time than is available. Thus, multigrid methods cycle between different resolution levels, performing some single-site updates at each level, in order to speed up computations and convergence to an accurate solution. There are several possible multigrid cycle regimes, the most common of them in applied mathematics and physics are known as the V and W cycles. The V cycle starts at the finest resolution level and progressively moves to coarser resolutions until it reaches the coarsest resolution level. After that it progressively moves to finer resolutions until it reaches the finest resolution level. The W cycle corresponds to the application of two successive V cycles. Other cycle regimes are possible, such as going only from coarser to finer levels analogously to the estimation of permeability fields in Chapter 16, or the grid can be dynamically adjusted, allowing the solver to move between different grids in an attempt to speed the solution without losing accuracy in the final result. This latter approach has been applied to statistical problems. For example, in the rejoinder to the main article, Besag et al. (1995, pp. 64–65) use this approach to allow their MCMC to move between scales when fitting Markov random field models to image data. In that context, multigrid methods become functionally equivalent to simulated sintering.

14

Metropolis-Coupled Methods

The methods of the previous chapter are useful for employing a multiscale approach for computational efficiency but are not as useful when one is faced with truly multiscale data. In contrast, the methods in this chapter are equally useful as a computational tool or for a full analysis of multiscale data. These methods can be seen as a generalization of simulated sintering, where instead of having a single chain moving between scales, multiple chains are run so that there is one (or more) chain at each scale at each point in time. These chains periodically exchange scales with each other, which allows information to move across scales but also ensures that we continue to sample at every scale. This eliminates some of the difficulties in posterior inference that can occur when trying to incorporate samples which are disjoint in time. In particular, simulated sintering produces a single MCMC run over the joint space of the parameters and the scale, but for any particular scale, there will only be MCMC samples at unconnected intervals.

14.1 Metropolis Coupling

The concept of Metropolis coupling was introduced by Geyer (1991) to improve mixing, although in the context of a standard MCMC run at a single scale (Barone et al., 2002). Coupling was later used in other contexts, such as diagnosing convergence (e.g., Johnson, 1998). For more context in one of the original intended uses of Metropolis coupling, see Section 14.3 below. In this chapter, we will follow the ideas of Higdon et al. (2002) in applying coupling to multiscale problems.

The basic idea of Metropolis coupling is to have chains running in parallel using standard MCMC updates (Gibbs sampling or Metropolis-Hastings techniques as appropriate). Periodically, these chains then propose to swap information in the sense of exchanging some or all of their parameter values. This step is done via the Metropolis-Hastings algorithm. We illustrate this here in the case of just two chains. Denote the parameter values of the first chain by θ_0 and those of the second chain by θ_1, where these could be vector-valued in multivariate problems. We use an additional subscript in parentheses to indicate the iteration number; e.g., $\theta_{1(1)}$ is the value of θ_1 at the first iteration, and $\theta_{1(4)}$ is its value at the fourth iteration. A schematic illustration of the process is

$$
\begin{array}{ccccccccccc}
\theta_{0(1)} & \xrightarrow{\text{MCMC}} & \theta_{0(2)} & \xrightarrow{\text{SWAP}} & \theta_{0(3)} & \xrightarrow{\text{MCMC}} & \theta_{0(4)} & \xrightarrow{\text{SWAP}} & \theta_{0(5)} & \xrightarrow{\text{MCMC}} & \cdots \\
\theta_{1(1)} & \xrightarrow{\text{MCMC}} & \theta_{1(2)} & & \theta_{1(3)} & \xrightarrow{\text{MCMC}} & \theta_{1(4)} & & \theta_{1(5)} & \xrightarrow{\text{MCMC}} & \cdots
\end{array}
$$

Here the arrows labeled $\xrightarrow{\text{MCMC}}$ are standard within-chain updates (one or more iterations), and the arrows labeled $\xrightarrow{\text{SWAP}}$ are updates that propose swapping information across chains. Technically, what is going on is that a single (yet complicated) multivariate chain is being run on the joint space of (θ_0, θ_1). The standard updates deal with θ_0 and θ_1 independently, as if two separate chains were being run. However, these updates can also be viewed as sequential updates in a single joint chain, where the two components happen to be independent. The swap steps then tie the two otherwise independent chains together.

To make this concrete, we take a simple example. Suppose we have data $\{y_1, \ldots, y_n\}$ which are a random sample from a normal distribution with unknown mean μ and variance σ^2. The likelihood is

$$
f(\mathbf{y}|\mu, \sigma^2) = (2\pi\sigma^2)^{-n/2} \exp\left\{ -\frac{1}{2\sigma^2} \sum_{i=1}^{n} (y_i - \mu)^2 \right\}.
$$

Denote the prior for the parameters as $\pi(\mu, \sigma^2)$, which could be the standard independent noninformative conjugate prior but could also be any other prior. One prior that could cause sampling issues would be $\pi(\mu, \sigma^2) = (\exp\{-(\mu - 3)^2 - (\mu + 3)^2\})/\sqrt{(2\pi)\sigma^2}$, which has μ and σ^2 a priori independent, puts a standard noninformative prior on σ, and puts a mixture prior on μ with highly separated masses at 3 and -3. This sort of prior could lead to a highly bimodal posterior (for example, if there aren't that many data points and they have a mean near zero). The posterior is $f(\mu, \sigma^2|\mathbf{y}) \propto f(\mathbf{y}|\mu, \sigma^2)\pi(\mu, \sigma^2)$, which could be sampled via Gibbs sampling for certain forms of π or else with Metropolis-Hastings algorithm in the general case. If one tried to use standard Metropolis-Hastings techniques with a bimodal posterior (such as the example above), then the Markov chain would be likely to get stuck in one of the two modes. One solution is to use a Metropolis-coupled chain. Start by creating two separate MCMC runs, one with (μ_0, σ_0^2) and the other with (μ_1, σ_1^2), and start them near separate modes of the posterior. Each chain can be updated independently in the beginning, with both using identical formulas for the likelihood and prior. Periodically, a swap step is made which proposes setting the new value for μ_0 (denoted μ_0^*) to be the current value of μ_1 and the new value for μ_1 (denoted μ_1^*) to be the current value for μ_0. This is a simple Metropolis proposal, which is accepted with probability

$$\frac{f(\mathbf{y}|\mu_0^*, \sigma_0^2)\pi(\mu_0^*, \sigma_0^2)f(\mathbf{y}|\mu_1^*, \sigma_1^2)\pi(\mu_1^*, \sigma_1^2)}{f(\mathbf{y}|\mu_0, \sigma_0^2)\pi(\mu_0, \sigma_0^2)f(\mathbf{y}|\mu_1, \sigma_1^2)\pi(\mu_1, \sigma_1^2)}$$

$$= \frac{f(\mathbf{y}|\mu_1, \sigma_0^2)\pi(\mu_1, \sigma_0^2)f(\mathbf{y}|\mu_0, \sigma_1^2)\pi(\mu_0, \sigma_1^2)}{f(\mathbf{y}|\mu_0, \sigma_0^2)\pi(\mu_0, \sigma_0^2)f(\mathbf{y}|\mu_1, \sigma_1^2)\pi(\mu_1, \sigma_1^2)} , \quad (14.1)$$

or with probability 1 if this ratio is larger than 1 (this condition is always implied if unstated). If the priors for μ and σ^2 are independent (i.e., $\pi(\mu, \sigma^2) = \pi(\mu)\pi(\sigma^2)$), then the swap acceptance probability (14.1) reduces to

$$\frac{f(\mathbf{y}|\mu_1, \sigma_0^2)f(\mathbf{y}|\mu_0, \sigma_1^2)}{f(\mathbf{y}|\mu_0, \sigma_0^2)f(\mathbf{y}|\mu_1, \sigma_1^2)}$$

because all the terms in the prior cancel. Note that the proposal is deterministic and reversible, so we do not need to include the Hastings term in the probability. If we now marginalize out μ_1 and σ_1^2, we get a posterior sample for μ_0 and σ_0^2 that has fully explored both modes of the posterior. Marginalizing is trivial, as we can simply discard all of the μ_1 and σ_1^2 values.

In this particular example, we swap only the μ values. If we had swapped both μ and σ, then the swap would always be accepted. But the resulting chain would be essentially the same as running two chains independently, as the likelihoods and priors are the same for each chain. Of course, this is merely an illustrative example. There are more efficient ways of exploring a multimodal posterior such as this. We have gone over these details to make the concept clear, and in the next section we apply this approach to the multiscale setting.

14.2 Multiscale Metropolis Coupling

Following Higdon et al. (2002), we now expand the original idea of Metropolis coupling to multiscale problems. We consider first just two scales, a coarse scale and a fine scale. Let $\boldsymbol{\theta}_0$ be the parameters at the coarse scale, with likelihood $f(\mathbf{y}_0|\boldsymbol{\theta}_0)$, and let $\boldsymbol{\theta}_1$ be the parameters at the fine scale, with corresponding likelihood $f(\mathbf{y}_1|\boldsymbol{\theta}_1)$. In some cases, such as the hydrology example in Chapter 16, the data will be the same at all scales (i.e., $\mathbf{y}_0 = \mathbf{y}_1$), and it is the parameters that are of interest at different scales. In other cases, we will actually observe different data at different scales, and in yet other cases, we may only observe data at the finest scale and may compute a coarser dataset from the fine data. The coarse and fine parameter spaces may also be the same or different, depending on the problem. Denote the respective priors by $\pi(\boldsymbol{\theta}_0)$ and $\pi(\boldsymbol{\theta}_1)$.

A pictorial representation of the resulting algorithm is

$$
\begin{array}{cccccccccc}
(\boldsymbol{\theta}_0)^1 & \xrightarrow{\text{MCMC}} & (\boldsymbol{\theta}_0)^2 & \xrightarrow{\text{SWAP}} & (\boldsymbol{\theta}_0)^3 & \xrightarrow{\text{MCMC}} & (\boldsymbol{\theta}_0)^4 & \xrightarrow{\text{SWAP}} & (\boldsymbol{\theta}_0)^5 & \xrightarrow{\text{MCMC}} \cdots \\
(\boldsymbol{\theta}_1)^1 & \xrightarrow{\text{MCMC}} & (\boldsymbol{\theta}_1)^2 & & (\boldsymbol{\theta}_1)^3 & \xrightarrow{\text{MCMC}} & (\boldsymbol{\theta}_1)^4 & & (\boldsymbol{\theta}_1)^5 & \xrightarrow{\text{MCMC}} \cdots
\end{array}
$$

Again, we create independent chains at the coarse and fine levels, and start with standard within-scale MCMC updates. Periodically we now propose to swap information between the two scales. The exact swapping mechanism will depend on the particular problem. The general idea is to move the information about the current fit at the coarse level to the fine level and vice versa. This may involve directly swapping the values of the parameters at the two scales if the parameter spaces are the same size and have the same interpretation (as with the convolution example below). Otherwise, a transformation of the parameters will be necessary to create a realistic proposal across the scales (as with the Markov random field models below). The probability of accepting the swap proposal is thus

$$\frac{f(\mathbf{y}_0|\boldsymbol{\theta}_0^*)\pi(\boldsymbol{\theta}_0^*)f(\mathbf{y}_1|\boldsymbol{\theta}_1^*)\pi(\boldsymbol{\theta}_1^*)q(\boldsymbol{\theta}_0,\boldsymbol{\theta}_1|\boldsymbol{\theta}_0^*,\boldsymbol{\theta}_1^*)}{f(\mathbf{y}_0|\boldsymbol{\theta}_0)\pi(\boldsymbol{\theta}_0)f(\mathbf{y}_1|\boldsymbol{\theta}_1)\pi(\boldsymbol{\theta}_1)q(\boldsymbol{\theta}_0^*,\boldsymbol{\theta}_1^*|\boldsymbol{\theta}_0,\boldsymbol{\theta}_1)}, \tag{14.2}$$

where $\boldsymbol{\theta}^*$ is the proposed value for each scale and $q(\boldsymbol{\theta}_0^*,\boldsymbol{\theta}_1^*|\boldsymbol{\theta}_0,\boldsymbol{\theta}_1)$ is the density function for the generation of the proposed values $(\boldsymbol{\theta}_0^*,\boldsymbol{\theta}_1^*)$ given the current values $(\boldsymbol{\theta}_0,\boldsymbol{\theta}_1)$. In the simple example of the previous section, this was a deterministic function, so its value was always 1. We must also take care to make sure that this proposal distribution q is reversible (as for all MCMC), so that the term in the denominator is always nonzero.

In many cases, the formula above can be simplified based on the particular form of the likelihood, prior, and swap proposal. In the next section, we illustrate this algorithm on the Fraser River data introduced in Chapter 3. The model used here is the Gaussian process convolution approach, described in more detail in Chapter 4. A single-resolution Gaussian process is fit at each scale. This approach has several key features that allow for simpler implementation.

A second class of models that are amenable to this multiscale approach are Markov random fields. Section 14.2.2 discusses the implementation of these models, and additional examples appear in Sections 15.4 and 16.3.

14.2.1 Swapping with Convolutions

One approach is to model the data with a Gaussian process, and to use convolutions to fit this model. The convolution concept was introduced in Chapter 4. Here we use a simple (single-scale) convolution at each scale. By putting the two datasets on comparable grids, the same kernel width and same latent process locations can be used, simplifying the swapping mechanism.

Working with the seasonally detrended log river data, the monthly flow averages are $\{y_{11},\ldots,y_{1n_1}\}$, and we also use quarterly averages $\{y_{01},\ldots,y_{0n_0}\}$. To keep the example simple, we will use only the first three years of the data, and hence $n_1 = 36$ and $n_0 = 12$. We consider $\mathbf{y}_1$ to be observed at the locations $\mathbf{s}_1 = \{1,2,\ldots,n_1\}$ and $\mathbf{y}_0$ to be observed at $\mathbf{s}_0 = \{2,5,8,\ldots,35\}$. Latent processes $\mathbf{w}_0$ and $\mathbf{w}_1$ are then defined on the grid $\mathbf{t} = \{-1,2,5,8,\ldots,35,38\}$, and these processes are convolved with Gaussian kernels k with standard deviation 3, $k(t_j - s_i) = \exp\{-(t_j - s_i)^2/18\}/\sqrt{18\pi}$. The same grid of $n_w = 14$ points and the same kernels are used on both scales. The fitted process is thus

$$\hat{y}_{ri} = \hat{y}_r(s_{ri}) = \sum_{j=1}^{n_w} k(t_j - s_{ri})w_{rj} \qquad (14.3)$$

for either the coarse scale ($r = c$) or the fine level ($r = f$). We take an *iid* Gaussian likelihood at each scale, with respective variances σ_0^2 and σ_1^2. However, it turns out it will be easier to work with precisions, which are the reciprocals of the variances, so we will use $\tau_0 = 1/\sigma_0^2$ and $\tau_1 = 1/\sigma_1^2$. During MCMC, the precision τ is updated via Gibbs sampling, as its complete conditional is a gamma distribution, while the variance σ^2 has a complete conditional that is an inverse-gamma and is thus a little more difficult to work with, and it would be a little bulkier in the already long equations that follow below. Thus, using precisions, the likelihoods at each scale are

$$f(\mathbf{y}_0|\mathbf{w}_0, \tau_0) = \left(\frac{\tau_0}{2\pi}\right)^{n_0/2} \exp\left\{-\frac{\tau_0}{2}\sum_{i=1}^{n_0}(y_{0i} - \hat{y}_{0i})^2\right\},$$

$$f(\mathbf{y}_1|\mathbf{w}_1, \tau_1) = \left(\frac{\tau_1}{2\pi}\right)^{n_1/2} \exp\left\{-\frac{\tau_1}{2}\sum_{i=1}^{n_1}(y_{1i} - \hat{y}_{1i})^2\right\}.$$

Note that the $\hat{y}$ terms are deterministic transformations of the appropriate $\mathbf{w}$, as given by Equation (14.3). Next, we put a prior on the latent w_i terms, each being *iid* Gaussian with mean zero and precision either λ_0 or λ_1, for the coarse and fine scales. Finally, we can either set these precisions to fixed values or can use a hierarchical structure and put priors on them. In a hierarchical setup, gamma priors will allow Gibbs sampling, so they are the obvious choice. One possibility is to use common hyperpriors for the scales,

$$\tau_0 \sim \Gamma(\alpha_\tau, \beta_\tau),$$
$$\tau_1 \sim \Gamma(\alpha_\tau, \beta_\tau),$$
$$\lambda_0 \sim \Gamma(\alpha_\lambda, \beta_\lambda), \qquad (14.4)$$
$$\lambda_1 \sim \Gamma(\alpha_\lambda, \beta_\lambda),$$

and then α_τ, α_λ, β_τ, and β_λ are constants that need to be specified. In this example, we use fairly weak priors (small α) with $\alpha_\tau = \alpha_\lambda = 1$, $\beta_\tau = 0.005$, and $\beta_\lambda = 1000$, i.e., τ_0 and τ_1 have exponential priors with mean 200. This specification encourages the process to fit the data with only some smoothing and allows large flexibility in the latent process.

With this setup, the swap maneuver is quite straightforward. We basically propose exchanging all values of $\mathbf{w}_0$ and $\mathbf{w}_1$. Both scales use a latent process on the same grid, so these latent values have a common interpretation.

The difference in the scales is from different discretizations of the convolved processes, both of which are in reality continuous processes. With such a swap proposal, the q term in Equation (14.2) is just 1. The τ terms are not part of the swap. If the same priors are used for the latent processes at both scales (i.e., $\lambda_0 = \lambda_1$), those terms also drop out of the equation, and thus the probability of accepting the swap is

$$\frac{f(\mathbf{y}_0|\mathbf{w}_1, \tau_0)f(\mathbf{y}_1|\mathbf{w}_0, \tau_1)}{f(\mathbf{y}_0|\mathbf{w}_0, \tau_0)f(\mathbf{y}_1|\mathbf{w}_1, \tau_1)}.$$

In some cases, these swaps may have a low probability of acceptance because the resulting fits are a little worse at both scales before they have a chance to settle down. It can then help to resample both τ terms as part of the swap proposal. The new proposal for τ is best chosen using the same Gibbs sampler that is being used for the standard τ updates. For example, using the hierarchical prior specified in Equation (14.4), the complete conditional distribution for τ_0 is a gamma distribution with parameters $\alpha_\tau + n_1/2$ and $\beta_\tau + SS_x/2$, where SS_x is the sum of squared differences between the observed data $\mathbf{y}$ and the fitted values $\hat{\mathbf{y}}$. However, this is not a Gibbs step because it is now just another piece of the multivariate swap, so it needs to be part of the sampling probability in the Metropolis-Hastings ratio, and it appears as the q terms below. Denoting the proposed values as τ^*, the resulting acceptance probability is now

$$\frac{f(\mathbf{y}_0|\mathbf{w}_1, \tau_0^*)\pi(\tau_0^*)f(\mathbf{y}_1|\mathbf{w}_0, \tau_1^*)\pi(\tau_1^*)q(\tau_0|\mathbf{w}_0)q(\tau_1|\mathbf{w}_1)}{f(\mathbf{y}_0|\mathbf{w}_0, \tau_0)\pi(\tau_0)f(\mathbf{y}_1|\mathbf{w}_1, \tau_1)\pi(\tau_1)q(\tau_0^*|\mathbf{w}_1)q(\tau_1^*|\mathbf{w}_0)}.$$

If a hierarchical prior (such as in Equation (14.4)) is used so that the latent processes have different hyperparameters λ_0 and λ_1, then these also need to be taken into account and can be redrawn as well. Again, it can be helpful to use a Gibbs-style draw to produce the proposal. This setup results in the acceptance probability

$$\frac{f(\mathbf{y}_0|\mathbf{w}_1, \tau_0^*)\pi(\tau_0^*)q(\tau_0|\mathbf{w}_0)f(\mathbf{y}_1|\mathbf{w}_0, \tau_1^*)\pi(\tau_1^*)q(\tau_1|\mathbf{w}_1)}{f(\mathbf{y}_0|\mathbf{w}_0, \tau_0)\pi(\tau_0)q(\tau_0^*|\mathbf{w}_1)f(\mathbf{y}_1|\mathbf{w}_1, \tau_1)\pi(\tau_1)q(\tau_1^*|\mathbf{w}_0)}$$

$$\times \quad \frac{\pi(\mathbf{w}_1|\lambda_0^*)\pi(\lambda_0^*)q(\lambda_0|\mathbf{w}_0)\pi(\mathbf{w}_0|\lambda_1^*)\pi(\lambda_1^*)q(\lambda_1|\mathbf{w}_1)}{\pi(\mathbf{w}_0|\lambda_0)\pi(\lambda_0)q(\lambda_0^*|\mathbf{w}_1)\pi(\mathbf{w}_1|\lambda_1)\pi(\lambda_1)q(\lambda_1^*|\mathbf{w}_0)}.$$

Note that the term $\pi(\mathbf{w}_1|\lambda_0^*)$ is the result of moving the latent process from the fine scale to the coarse scale, resampling λ_0 from the distribution $q(\lambda_0^*|\mathbf{w}_1)$ to get a proposed value λ_0^*, and then evaluating the prior for the latent process (on the coarse scale, with the values obtained from the fine scale, and using this newly proposed λ^*).

For the Fraser River example with the hierarchical prior above, after cancelling out various normalizing constants, the swap acceptance ratio is

$$\frac{(\tau_0^*)^{n_0/2}\exp\left\{-\frac{\tau_0^*}{2}SS_{y0}^*\right\}(\tau_0^*)^{\alpha_\tau-1}\exp\{-\beta_\tau\tau_0^*\}(\tau_0)^{(\alpha_\tau+n_0/2)-1}\exp\{-(\beta_\tau+SS_{y0}/2)\tau_0\}}{(\tau_0)^{n_0/2}\exp\left\{-\frac{\tau_0}{2}SS_{y0}\right\}(\tau_0)^{\alpha_\tau-1}\exp\{-\beta_\tau\tau_0\}(\tau_0^*)^{(\alpha_\tau+n_0/2)-1}\exp\{-(\beta_\tau+SS_{y0}^*/2)\tau_0^*\}}$$

$$\times\frac{(\tau_1^*)^{n_1/2}\exp\left\{-\frac{\tau_1^*}{2}SS_{y1}^*\right\}(\tau_1^*)^{\alpha_\tau-1}\exp\{-\beta_\tau\tau_1^*\}(\tau_1)^{(\alpha_\tau+n_1/2)-1}\exp\{-(\beta_\tau+SS_{y1}^*/2)\tau_1\}}{(\tau_1)^{n_1/2}\exp\left\{-\frac{\tau_1}{2}SS_{y1}\right\}(\tau_1)^{\alpha_\tau-1}\exp\{-\beta_\tau\tau_1\}(\tau_1^*)^{(\alpha_\tau+n_1/2)-1}\exp\{-(\beta_\tau+SS_{y1}^*/2)\tau_1^*\}}$$

$$\times\frac{(\lambda_0^*)^{n_w/2}\exp\left\{-\frac{\lambda_0^*}{2}SS_{w1}\right\}(\lambda_0^*)^{\alpha_\lambda-1}\exp\{-\beta_\lambda\lambda_0^*\}(\lambda_0)^{(\alpha_\lambda+n_w/2)-1}\exp\{-(\beta_\lambda+SS_{w0}/2)\lambda_0\}}{(\lambda_0)^{n_w/2}\exp\left\{-\frac{\lambda_0}{2}SS_{w0}\right\}(\lambda_0)^{\alpha_\lambda-1}\exp\{-\beta_\lambda\lambda_0\}(\lambda_0^*)^{(\alpha_\lambda+n_w/2)-1}\exp\{-(\beta_\lambda+SS_{w1}/2)\lambda_0^*\}}$$

$$\times\frac{(\lambda_1^*)^{n_w/2}\exp\left\{-\frac{\lambda_1^*}{2}SS_{w0}\right\}(\lambda_1^*)^{\alpha_\lambda-1}\exp\{-\beta_\lambda\lambda_1^*\}(\lambda_1)^{(\alpha_\lambda+n_w/2)-1}\exp\{-(\beta_\lambda+SS_{w1}/2)\lambda_1\}}{(\lambda_1)^{n_w/2}\exp\left\{-\frac{\lambda_1}{2}SS_{w1}\right\}(\lambda_1)^{\alpha_\lambda-1}\exp\{-\beta_\lambda\lambda_1\}(\lambda_1^*)^{(\alpha_\lambda+n_w/2)-1}\exp\{-(\beta_\lambda+SS_{w0}/2)\lambda_1^*\}},$$

where $SS_{y0} = \sum_{i=1}^{n_0}(y_{0i}-\hat{y}_{0i})^2$ and $SS_{y0}^* = \sum_{i=1}^{n_0}(y_{0i}-\hat{y}_{0i}^*)^2$, and $\hat{y}_0$ is given by Equation (14.3) and is a function of $\mathbf{w}_0$, while $\hat{y}_0^*$ is a function of $\mathbf{w}_1$. Similarly, $SS_{w0} = \sum_{j=1}^{n_w} w_{0j}^2$. The corresponding quantities for the fine scale are defined analogously. Careful inspection shows that this ratio is exactly equal to 1 (because in each case $q \propto f\pi$), so that these swaps are always accepted, but it is displayed as above so the reader can easily see where all of the terms come from, and how one would modify it for a different setup where it may not be so easy to sample from the complete conditional, such as with the Markov random fields in the next section. But, in the present case, the swaps are accepted with probability 1, improving the mixing of the Markov chain.

Figure 14.1 shows the results of running our R code (available at http://www.ams.ucsc.edu/~herbie/multiscale) for Metropolis coupling on the first three complete years of the detrended Fraser River data. The data are shown as the circles, with open circles for the original (fine scale) data and solid circles for the quarterly averages (coarse scale). Vertical dashed lines show the locations of the latent process. The solid line is the posterior mean fitted curve at the fine scale, and the middle dash-dot line is the fitted curve at the coarse scale. Both curves are quite similar; 95% credible intervals are shown as the dashed (fine scale) and dotted lines (coarse scale). This problem is not too complex, and the interval estimates are also similar at both scales. Note that, under the Bayesian approach, uncertainty estimates are automatically available as a result of having fit the model with MCMC, and are available at all scales.

One of the key benefits of Metropolis coupling is improved mixing and posterior exploration. One possible measure of mixing is the autocorrelation of elements of the Markov chain or derived quantities (Sokal, 1987). This problem is fairly simple, so we won't be able to show any drastic improvements. A more dramatic improvement appears in the hydrology example in Chapter 16. Here we show a more modest but still noticeable improvement. Figure 14.2 shows the autocorrelation plots for one of the latent process values at the fine (original) scale, with a single-resolution run (at the fine scale only) on the left and the Metropolis-coupled run on the right. Notice how the autocorrelations die out more quickly using Metropolis coupling, in that the correlation at lag

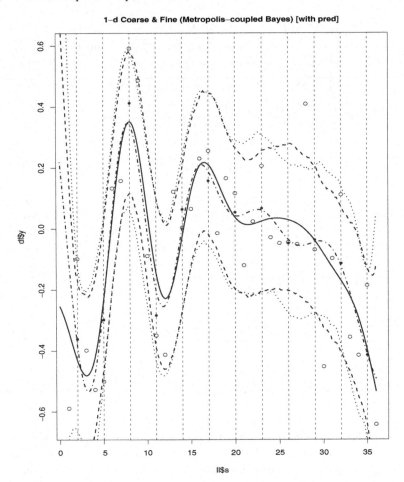

Fig. 14.1. Fitted values and confidence bands for the Fraser River data. Open circles are the original data, solid circles are quarterly averages, the solid line is the posterior mean fitted curve at the fine scale, and the middle dashed line is the fitted curve at the coarse scale; 95% credible intervals are shown with dashed lines for the fine scale and dotted lines for the coarse scale.

one is smaller, and the correlation at lag two is nearly zero and well below the standard error line. We could claim that the correlation range has been reduced by one-half, although this is really just from two lags to one. While it is true that autocorrelation is not necessarily the optimal measure of mixing, it is an easily obtainable and relatively intuitive measure. The point here is that there is a reduction in autocorrelation, and that this reduction can be pronounced in more complex situations.

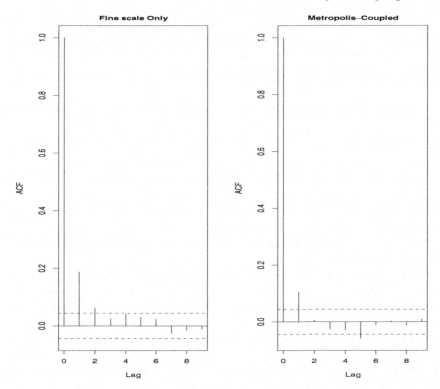

Fig. 14.2. Autocorrelation plots for a latent value.

14.2.2 Swapping with Autoregressive Processes

For spatial data, an alternative to a Gaussian process model is a Markov
random field (MRF), and these were introduced in Section 2.1. While the
Gaussian process models were easy to deal with at different scales because the
underlying parameters were the same (specifying a continuous process that
was merely discretized at different scales but always had the same parameter
dimension), an MRF specifies the process over a grid and thus must specify
a value for each grid cell. Thus, if a fine scale has twice as many grid cells
in each dimension, then the number of parameters will increase by a factor
of 2 for each dimension. For the sake of illustration, we will discuss a two-
dimensional process here, as the examples that appear in Sections 15.4 and
16.3 are two-dimensional. So if the coarse scale is a 32×32 grid and the fine
scale is a 64×64 grid, there are four times as many parameters at the fine
scale than at the coarse scale, and thus there is a four-to-one correspondence
that must be dealt with when swapping across scales.

Here the MRF could be either proper as in Section 10.1, or it could be intrinsic and thus not strictly stationary (because its mean level is not defined).

Because the examples use intrinsic fields, as well as the fact that intrinsic fields result in simpler notation, we focus on the intrinsic case in the rest of this section. The basic model for a first-order intrinsic Gaussian MRF at a single scale is defined as values x_j over a prespecified grid,

$$\pi(\mathbf{x}|\beta) \ \propto \ \beta^{\frac{m}{2}} \exp\left\{-\frac{1}{2}\beta \sum_{i \sim j}(x_i - x_j)^2\right\} \ = \ \beta^{\frac{m}{2}} \exp\left\{-\frac{1}{2}\beta \mathbf{x}^T \mathbf{H} \mathbf{x}\right\},$$

(14.5)

where β is a smoothing parameter, m is the total number of grid cells, and $i \sim j$ specifies that grid cells i and j are neighbors. In two dimensions, interior cells will have four neighbors, edge cells three, and corner cells two. In one dimension, a cell would have only one or two neighbors. In matrix notation, the precision matrix $\mathbf{H}$ contains off-diagonal entries $h_{ij} = -1$ if pixel i is adjacent to pixel j and 0 otherwise, and diagonal elements h_{ii} equal to the number of neighbors of pixel i.

In order to create the correspondence between scales, we work with the means of equivalent blocks. Thus, when moving between scales, we consider a joint move between four fine cells and the one corresponding coarse cell (in the two-dimensional case), and we keep the mean of the four fine cells equal to the mean of the corresponding coarse cell (which is just its value). This formulation makes the swap proposal from the fine to the coarse levels a deterministic proposal, $x_{0i}^* = \frac{1}{4}\sum_{j \in J_i} x_{1j}$, where J_i is the set of indices of the four fine cells that correspond to the ith coarse cell. The full proposal is then the collection of the subproposals for each of the individual cells; i.e., for each cell at the coarse scale, its proposed value is found by averaging the current values of the four corresponding fine cells.

One possible proposal for going from the coarse to the fine scale would be to deterministically set all four corresponding fine cells to be the value of the current coarse cell. While conceptually simple, this tends not to work as well in practice because the resulting fine scale has a very blocky behavior and is unlikely to be consistent with the MRF prior at the fine scale (which expects the variability between adjacent cells to be at a more constant level, rather than some neighbors taking identical values and some taking very different values). If the proposal is not consistent with the prior, then the prior term in the Metropolis-Hastings acceptance probability will be very small and the proposal will be rejected. In order to rectify this situation, a more complex and more smooth proposal is needed.

The approach of Higdon et al. (2002) is to sequentially draw fine cell proposals from the prior distribution conditioned on the blocks of four having a mean matching the current corresponding coarse cell. These draws are done block-by-block, with the other blocks being held fixed, either at temporary values before they are updated, or using the updated values after they have been drawn. For the MRF specification, conditioned on the other blocks, the draw for the four cells in a block is a multivariate normal restricted to have a particular mean. In the two-dimensional case, this would be a degenerate four-dimensional draw, or equivalently a three-dimensional multivariate normal. To be specific, the steps for creating the fine proposal are as follows:

1. Set the temporary value of each cell x_{1j}^* in a fine block of four equal to the corresponding coarse value x_{0i}.
2. Repeat step 1 for all of the cells at the fine level.
3. For each fine block, draw a random sample from the MRF prior distribution conditioned on the block having the fixed mean, and conditioned on the currently proposed values of the neighboring cells from all neighboring blocks.
4. Repeat step 3 for each of the fine blocks, so that the original temporary values are sequentially replaced by these multivariate draws.

This process will create an entire fine field $\mathbf{x}_1^*$.

As with the Gaussian process specification, it is usually also helpful to include new proposals for the precision parameters β_1 and β_0, and these can be drawn from their complete conditional distributions using the proposed values $\mathbf{x}_1^*$ and $\mathbf{x}_0^*$. This joint proposal of $(\mathbf{x}_1^*, \mathbf{x}_0^*, \beta_1^*, \beta_0^*)$ is then accepted or rejected according to the standard Metropolis-Hastings probability of

$$\frac{f(\mathbf{y}_0|\mathbf{x}_0^*, \beta_0^*)\pi(\mathbf{x}_0^*|\beta_0^*)\pi(\beta_0^*)f(\mathbf{y}_1|\mathbf{x}_1^*, \beta_1^*)\pi(\mathbf{x}_1^*|\beta_1^*)\pi(\beta_1^*)q(\mathbf{x}_1|\mathbf{x}_0^*, \beta_1^*)q(\beta_0|\mathbf{x}_0)q(\beta_1|\mathbf{x}_1)}{f(\mathbf{y}_0|\mathbf{x}_0, \beta_0)\pi(\mathbf{x}_0|\beta_0)\pi(\beta_0)f(\mathbf{y}_1|\mathbf{x}_1, \beta_1)\pi(\mathbf{x}_1|\beta_1)\pi(\beta_1)q(\mathbf{x}_1^*|\mathbf{x}_0, \beta_1)q(\beta_0^*|\mathbf{x}_0^*)q(\beta_1^*|\mathbf{x}_1^*)}.$$

Because the coarse proposal is a deterministic function of the current fine values, the q terms are equal to 1 and are omitted. If the likelihood or prior are more complex, further parameters can be included as in the previous section. More details on this procedure can be found in Higdon et al. (2002). Examples can be found in Chapter 17 for the SPECT data, in the context of Metropolis-coupled swaps and genetic-algorithm style swaps, and in Section 16.3 in the hydrology case study.

14.3 Sequential Parallel Tempering

Similar algorithms have been developed independently under the name of sequential parallel tempering (Liang, 2003). In the next paragraph, we discuss parallel tempering, before moving on to the sequential version. As a side note, in fitting with the discussion of this chapter and the previous one, perhaps "parallel sintering" would have been a more descriptive name than sequential parallel tempering.

Parallel tempering (Geyer, 1991) extends the ideas of simulated tempering (from Section 13.2) to use multiple Markov chains simultaneously. The tempering ladder of functions at m different temperatures is set up, and instead of running a single chain that moves between the temperatures via an auxiliary variable, one chain is run at each temperature. Thus again one has a Markovian system over the joint space of the parameters and the temperature scale, but in this case there are m nominally independent chains running in parallel. The system of chains alternates between independent within-temperature updates and Metropolis-coupled updates across temperatures. These Metropolis-coupled updates are completely analogous to those described earlier in this chapter for moving between scales. Through this coupling, mixing is often greatly improved, allowing escapes from local maxima and better posterior exploration. As the chains are asymptotically independent, if one is only interested in the coolest distribution, one can just take the corresponding chain, discarding the others, and base posterior inference on the samples from that chain.

Sequential parallel tempering (Liang, 2003) goes one step further, combining multigrid/sintering methods with parallel tempering. The idea is to define a sequence of spaces of reduced dimension $\{\mathcal{X}_i\}$ such that $\dim(\mathcal{X}_1) > \dim(\mathcal{X}_2) > \ldots > \dim(\mathcal{X}_m)$. Choosing the $\mathcal{X}_i$ to be different scales is the obvious application here, although the algorithm is more broadly applicable (for example, dimension reduction via marginalization). In the context of multiscale modeling, by choosing the $\mathcal{X}_i$ to be different scales, the algorithm is functionally equivalent to the Metropolis-coupled methods previously discussed in this chapter.

14.4 Extensions

The swapping methods in this chapter are quite flexible, being readily applicable to a wide variety of situations. While the examples here were one-dimensional cases, the approach is equally useful in higher dimensions. Higdon et al. (2002) use Markov random fields, which is a generalization of the autoregressive model. Higdon et al. (2003) look at convolution models in higher dimensions.

It is also straightforward to extend this approach to more than two scales. An initially independent chain is set up at each scale, and then swaps are proposed between chains. When dealing with more than two scales, it is usually best to restrict swap proposals to adjacent scales, as those scales farther apart will tend to lead to low acceptance probabilities.

While the Metropolis-coupled algorithms were introduced with only a single chain at each scale, one could just as easily run multiple chains at each scale, as with parallel tempering or sequential parallel tempering. Having more such chains can be particularly beneficial in multiscale multimodal problems, just as with single-scale multimodal problems. The main restriction here is computer time.

As a final note, these algorithms are ideal for parallel computing environments. The chains are run mostly independently, with only occasional swapping interactions. The amount of information that has to be passed between processes is fairly small, usually just the set of relevant parameter values. Most of the computational work can be done by the processors independently, leading to a highly efficient use of a parallel environment.

15

Genetic Algorithms

This chapter starts by reviewing the key ideas of genetic algorithms and then demonstrates how to use them in a multiscale setting. The direct extension to multiscale modeling is useful for maximization. However, these multiscale ideas can also be extended to the Bayesian framework in combination with Markov chain Monte Carlo ideas. The resulting approach can then be seen as a generalization of the Metropolis-coupled methods of the previous chapter. This combination is a powerful method for fully Bayesian multiscale modeling, particularly in a parallel computing environment. Compared with the methods in the previous two chapters, the key extension of this chapter is to allow the swapping across scales (or "temperatures") of only part of the information in the chains instead of requiring the entire current state of the chain to move between scales. This adaptation can improve the probability of accepting a swap in complicated problems and thus further improve the mixing of the chains.

15.1 The Basics of Genetic Algorithms

Genetic algorithms were developed as a general maximization technique that was inspired by biological evolution. Maximization of a function with a higher-dimensional input parameter vector can be quite difficult when the function has many local maxima (i.e., it is multimodal). Genetic algorithms have been found to be one of the best methods for dealing with such complicated problems.

The concept in nature is that well-adapted organisms are more likely to survive and pass their genes on to the next generation. In each generation, there is the possibility of some random genetic mutations, and in the process of reproduction there is mixing and redistribution of genetic material. Over time, the genes which result in traits that are helpful for survival and reproduction (genes with high "fitness") will tend to dominate. This concept is then applied to the numerical maximization setting by thinking of a vector of parameter values as a chromosome of genes and by defining genetic fitness as the result of evaluating the function at those parameter values. A collection of these chromosomes is randomly mutated and mixed via transformations analogous to biological ones, as explained below. The collection is then culled, with those of higher fitness being more likely to be kept. This process is repeated and will converge to one or more local maxima, often including the global maximum even in difficult multimodal settings.

Early versions of the genetic algorithm (Holland, 1975) dealt with discrete parameter spaces with a relatively small number of possibilities, analogous to the biological setting where there are only four genetic bases. Later versions generalized this approach to real-valued parameters, which is also the primary interest for this book. Here we outline the basics of such algorithms. For more details, the reader is referred to Chatterjee et al. (1996) and the references therein, which provide an overview of the concepts as well as descriptions of applications in the field of statistics.

Suppose we wish to maximize some function $g(\mathbf{v})$, where $\mathbf{v}$ is a vector $(v_1, \ldots, v_R)$ and R is the number of parameters. $g(\mathbf{v})$ is thus taken as the measure of fitness. To initialize the algorithm, we need to create a population of solutions (analogous to chromosomes) each of which is represented as a vector of length R. For example, a population of solutions $\mathbf{v}_{(0)}^1, \ldots, \mathbf{v}_{(0)}^M$ could be created by randomly selecting M vectors of length R from some distribution. Here, the subscript in parentheses denotes the iteration number within the algorithm. Ordinarily, R is determined by the problem, but M, the number of solutions, may be chosen by the user. One cycle of the algorithm is comprised of three steps based on the biological analogy: selection, crossover, and mutation.

Selection: Mimicking biological natural selection, in the selection step, the population of solutions is culled to potentially remove poorer solutions while generally allowing better solutions to remain. Recall that the goal of the genetic algorithm is to maximize the fitness $g(\mathbf{v})$. The selection step proceeds by drawing M vectors with replacement from the current population of solutions to form a new population. If $g(\cdot)$ is nonnegative, solution vectors are selected with probabilities proportional to their fitnesses, $g(\mathbf{v})$. Thus, for one solution $\mathbf{v}_{(t)}^i$, its probability of being selected is $g(\mathbf{v}_{(t)}^i)/\sum_{j=1}^M g(\mathbf{v}_{(t)}^j)$. In this way, solutions with a higher fitness have a larger probability of being selected and moving through to the next cycle. If $g(\cdot)$ is not a nonnegative function, then it is typically transformed to be so, such as through exponentiating, $\exp(g)$. This one-to-one monotone transformation is then used as the fitness function.

Crossover: The crossover step is drawn from the idea of pairs of chromosomes swapping genetic material by having them exchange segments which are located at the same positions on both chromosomes. To perform a crossover, several pairs of solutions (up to $M/2$ pairs) are selected and some of their values are traded. It has been recommended that for each pair, the probability of performing a trade should be tuned to be between 0.6 and 0.95 (see Bäck, 1993, and the references therein). If a pair is selected to perform a trade, then one of several types of crossovers may be performed. In a one-point crossover, a single element is chosen randomly from $\{1, \ldots, R-1\}$ and all elements after the chosen element are swapped. The following diagram depicts a one-point crossover between $\mathbf{v}^1_{(t)}$ and $\mathbf{v}^2_{(t)}$ when the element z^* is chosen. The subscript denoting the iteration, (t), is suppressed within the expanded vectors for clarity.

$$\mathbf{v}^1_{(t)} = \{v^1_1, \ldots, v^1_{z^*}, v^1_{z^*+1}, \ldots, v^1_R\} \xrightarrow[\text{Swap}]{\text{Partial}} \mathbf{v}^1_{(t+1)} = \{v^1_1, \ldots, v^1_{z^*}, v^2_{z^*+1}, \ldots, v^2_R\}$$
$$\mathbf{v}^2_{(t)} = \{v^2_1, \ldots, v^2_{z^*}, v^2_{z^*+1}, \ldots, v^2_R\} \quad\quad \mathbf{v}^2_{(t+1)} = \{v^2_1, \ldots, v^2_{z^*}, v^1_{z^*+1}, \ldots, v^1_R\}$$

This particular style of crossover is called a single-point crossover. Some additional types of crossovers are discussed below.

Mutation: Following the random mutations that occur in nature, the mutation step randomly perturbs each element of each vector in the population of solutions with some small mutation probability. For instance, using a probability of mutation 0.01, in each cycle on average $0.01 \times M \times R$ elements would be altered. When working with real-valued vectors, mutation is typically performed by adding independent Gaussian noise to the selected elements.

These three steps are repeated until a prespecified measure of convergence is satisfied. Typical convergence criteria include: no change in maximum fitness over multiple iterations; the most fit solution obtains a fitness above a threshold level; or the number of iterations reaches a preestablished limit.

The crossover step as described above is for a single-point crossover, as directly inspired by genetics, where a single element is chosen and the solutions "cross over" at that point, swapping their remaining sections. A number of alternative styles of crossovers have been proposed; for example:

k-point crossover: k elements are chosen and segments between the chosen elements alternate between swapping and not swapping.

Uniform crossover: A swapping probability is chosen, and then each element is swapped individually with this probability.

Snooker crossover: Introduced by Liang and Wong (2001), this crossover move is intended to improve mixing by moving along directions of two solutions rather than merely exchanging solutions, and is based on the snooker algorithm (Roberts and Gilks, 1994). The steps of the crossover are as follows.

1. Select one solution, $\mathbf{v}^i$, chosen with equal probability from the population.

2. Select another solution, $\mathbf{v}^j$, from the remaining population with probabilities proportional to the fitness of the solutions.
3. Define $\mathbf{e} = (\mathbf{v}^j - \mathbf{v}^i)/\|\mathbf{v}^j - \mathbf{v}^i\|$ and then sample $u \in (-\infty, \infty)$ from the density $f(u) \propto |u|^{R-1} \pi(\mathbf{v}^j + u\mathbf{e})$, where π is the prior for the parameters.
4. Let $\tilde{\mathbf{v}} = \mathbf{v}^j + u\,\mathbf{e}$ and replace $\mathbf{v}^i$ with $\tilde{\mathbf{v}}$ in the population of solutions.

For the rest of this chapter, we will concentrate on single-point crossovers (for their simplicity) and uniform crossovers (as we have found they most help improve mixing in complex problems).

Several modifications to the basic genetic algorithm have been proposed to speed up convergence. One that we have found helpful is the elitist strategy (DeJong, 1975), where the solution with the highest fitness value is automatically retained after the mutation step; i.e., if the crossover and mutation steps cause the previously best solution to be worse, and no better solution arises, then the element with lowest fitness is immediately discarded and replaced by the previously best solution (so that it remains as the best solution).

15.2 Multiscale Genetic Algorithms

The form of the algorithm described above is suited only to solving problems for which all considered solutions have the same dimension, which is true in most problems but not generally true in multiscale problems. An early attempt to overcome the fixed solution length limitation was by Goldberg et al. (1989), who introduced a variable string length genetic algorithm. Their algorithm allows for solution representations of varying sizes in order to facilitate maximization. Although this algorithm only solves problems where the final solution is of a fixed dimension, it does demonstrate the utility of allowing different solution representations within the algorithm. The algorithm we describe here appeared in Holloman (2002) and Holloman et al. (2006) and was developed to extend the original genetic algorithm by making use of related solutions of differing dimensions.

The basic issue is that in order to perform crossover style swaps, it is necessary that the elements being swapped have a common interpretation. If that is not the case, then the resulting entity will not make any sense. For example, consider a time series on a daily scale and the same series on a weekly scale. A set of fitted values at the daily scale cannot be combined via a crossover with a set of fitted values at the weekly scale as the resulting combination would be a mix of daily and weekly values and not have any useful interpretation. What needs to be done is to establish a subset of elements of the solutions that share a common interpretation, so that these can be swapped. In the case of daily and weekly fitted values, the set of daily values could be reparameterized as the weekly average fitted value and the daily fitted differences from the weekly mean. Now both scales have weekly fitted values, and these can be swapped via any of the crossover moves described above.

To be more precise, we choose a parameterization such that we can partition each solution vector into two parts, $\mathbf{v} = (\boldsymbol{\phi}, \boldsymbol{\lambda})$, where $\boldsymbol{\phi}$ has a common interpretation across all scales and $\boldsymbol{\lambda}$ encodes all of the scale-specific information. An important factor is that $\boldsymbol{\phi}$ has the same dimension, C, regardless of the scale i. So to continue the example of daily and weekly values, $\boldsymbol{\phi}$ would be the weekly values or weekly averages, and $\boldsymbol{\lambda}$ might be empty for the weekly series, while for the daily series it would contain the differences between the daily values and the weekly averages. Thus one could swap the $\boldsymbol{\phi}$ values for the two scales and the resulting solutions would still make sense.

As with the previous chapter, the likelihood function must be specified at each scale, and then the joint likelihood is taken as their product. In some cases, the likelihoods may have the same dimensions across all scales, but in many cases, moving between scales changes both the number of available observations and the number of parameters in the model. When the dimension changes, it will be necessary to pay more attention to some of the details in the algorithm, as the normalizing constants can have a large effect.

To initialize the algorithm, we choose $M \geq 2 \times I$ random solution vectors, $\mathbf{v}_{(0)}^1, \ldots, \mathbf{v}_{(0)}^M$, where $\mathbf{v}_{(0)}^m = (\boldsymbol{\phi}_{(0)}^m, \boldsymbol{\lambda}_{(0)}^m, i_{(0)}^m)$ for $m = 1, \ldots, M$. The last parameter (i) in this solution vector specifies the resolution (and portion of the full likelihood) to which this solution vector corresponds. We recommend having at least two solution vectors for each resolution, though this is not strictly necessary. The algorithm often works better with even more solution vectors at each resolution, but having two can be a reasonable compromise between algorithm efficiency and computational resources. Once initialized, the algorithm follows selection, crossover, and mutation steps similar to those from before but modified to deal with the possibly differing dimensions of the parameters across scales.

One important distinction is in the selection step. Traditionally, selection is across all solution vectors. However, in the multiscale setting, the likelihoods may not be the same at different scales, and so one would not want to blindly compare fitness via different likelihoods. Instead, it is better to apply the selection step to each scale separately. Within a particular scale, the likelihood function is the same for all solutions, and so it can be used as a measure of fitness. Recall that the likelihoods for individual scales are conditionally independent, so implementation is straightforward.

The key change is in the crossover step, as this step would normally destroy the solutions when crossing between scales if not properly modified. Thus the crossover is modified so that only the $\boldsymbol{\phi}$ parts of the solutions are involved, and the $\boldsymbol{\lambda}$ parts are left unchanged by the crossover step. As with the standard crossover moves, the new solutions are not guaranteed to be good solutions, but at least they are legitimate solution vectors. Because all of the $\boldsymbol{\phi}$ vectors are the same length and have the same general meaning, this treatment ensures that meaningful solutions are produced by the crossover moves. Alternative crossovers are possible for strings of varying lengths (Goldberg et al., 1989); however, the crossover described here takes full advantage of the common interpretation of ϕ vectors on different scales. The steps of the full multiscale algorithm are the following.

Selection: Each of the I scales is dealt with separately. For a particular scale, the scale-specific likelihood serves as the fitness function for the $n_{(t)}^{(i)}$ solution vectors at that scale (identified by their last element). Now resample $n_{(t)}^{(i)}$ of these solutions with replacement with probabilities proportional to their likelihoods, $L^{(i)}(\phi^{(i)}, \lambda^{(i)} \mid y^{(i)})$, to form a new population at this level. This procedure is repeated for each $i \in \{1, \ldots, I\}$.

Crossover: The crossover step is modified from the ordinary genetic algorithm since the lengths of the vectors $v_{(t)}^1, \ldots, v_{(t)}^M$ may not all be the same. Instead we perform crossovers (of any type, as described in Section 15.1) on the portion of the vectors that are of the same length, the ϕ vectors. The last part of each vector, (λ, i), is not involved in the crossover. For example, a hypothetical one point crossover between two vectors is shown in the following diagram, where the subscript denoting the iteration, (t), is suppressed within the expanded vectors for clarity.

$$v_{(t)}^1 = \{\phi_1^1, \ldots, \phi_{z*}^1, \phi_{z*+1}^1, \ldots, \phi_C^1, \lambda^1, i^1\}$$
$$v_{(t)}^2 = \{\phi_1^2, \ldots, \phi_{z*}^2, \phi_{z*+1}^2, \ldots, \phi_C^2, \lambda^2, i^2\}$$

$$\overset{\text{Swap}}{\longrightarrow} \begin{array}{l} \{\phi_1^1, \ldots, \phi_{z*}^1, \phi_{z*+1}^2, \ldots, \phi_C^2, \lambda^1, i^1\} \\ \{\phi_1^2, \ldots, \phi_{z*}^2, \phi_{z*+1}^1, \ldots, \phi_C^1, \lambda^2, i^2\} \end{array}$$

Mutation: The mutation step remains basically unchanged from the standard genetic algorithm. Each element has a small probability of being perturbed, typically by adding random noise (for the case of real-valued parameters). It is not necessary to consider mutating the value of the scale parameter, i, because of the crossover moves, but it can be done and might speed up convergence in some cases. If i is mutated, one may want to take care that at least one solution is allowed to remain at each resolution in the solution space. Changing the value of i would also require adjusting λ appropriately.

Note that it is through the multiscale crossovers that information is shared across resolutions. These crossovers can be seen as a generalization of the Metropolis-coupled swaps of the previous chapter. While the Metropolis-coupled swaps exchanged complete parameter vectors, the genetic algorithm crossovers allow swapping of only pieces of the solutions through any of the crossover moves previously described, such as the single-point, k-point, uniform, or snooker crossovers.

As with ordinary genetic algorithms, using an elitist strategy that saves the best solution of interest from being destroyed through crossover or mutation may increase the speed of convergence (DeJong, 1975). In the multiscale case, one may want to retain the best solution at each scale rather than merely the best overall solution.

15.3 Multiscale Genetic Algorithm-Style MCMC

In the Bayesian paradigm, one is usually interested in estimating a posterior distribution for the parameters rather than simply finding an optimal point estimate. In particular, if the full parameter distribution is known, then the estimation of and accounting for uncertainty is straightforward. Here we retain the formulation of the likelihood and decomposition of the parameter vector $\psi = (\phi, \lambda)$ as in the previous section. To these we must add a prior distribution for the parameters at each scale $\psi^{(i)} = (\phi^{(i)}, \lambda^{(i)})$, which we denote $\pi^{(i)}(\psi^{(i)}) = \pi^{(i)}(\phi^{(i)}, \lambda^{(i)})$. The resulting posterior distribution for each scale (up to a normalizing constant) is

$$\pi^{(i)}\left(\psi^{(i)} \mid y^{(i)}\right) = \pi^{(i)}\left(\phi^{(i)}, \lambda^{(i)} \mid y^{(i)}\right) \propto L\left(\psi^{(i)} \mid y^{(i)}\right) \pi^{(i)}\left(\phi^{(i)}, \lambda^{(i)}\right).$$

As before, once the posterior is defined on a single scale, generalizing to multiple resolutions is simple. The full posterior is simply the product of the posteriors from each scale. To sample from the posterior in complex problems, the typical approach is to use Markov chain Monte Carlo (MCMC). The next section discusses combining genetic algorithms and MCMC. Then the key multiresolution adaptation is described in the following section. This algorithm can take advantage of the multiresolution nature of the problem to efficiently explore the parameter space and draw samples from the posterior.

As an aside, we note that this framework extends to more complex situations where one could have a parameter vector $\psi^{(i)}$ that was a non-deterministic transformation of $(\phi^{(i)}, \lambda^{(i)})$, in which case we must add another term, $\pi^{(i)}(\psi^{(i)} \mid \phi^{(i)}, \lambda^{(i)})$, to the right-hand side for the distribution induced by the probabilistic transformation, and the posterior will be defined over $\psi^{(i)}$, $\phi^{(i)}$, and $\lambda^{(i)}$. In addition, the term $\pi^{(i)}(\psi^{(i)} \mid \phi^{(i)}, \lambda^{(i)})$ will need to be included when appropriate in expressions in the following sections. Some further comments are available in Holloman et al. (2006).

15.3.1 Genetic Algorithms and MCMC

Just as optimization can be difficult in complex multimodal problems, MCMC can similarly be troubled by complex multimodal posterior distributions. The chain can easily become stuck in a local mode and fail to explore the whole space, never visiting other modes and possibly never visiting the most important parts of the space. As genetic algorithms are one of the more powerful optimization methods in the presence of multimodality, it makes sense to try to incorporate some of their features into MCMC to improve mixing and posterior exploration.

Holmes and Mallick (1998) suggest an MCMC algorithm, the Parallel Adaptive Metropolis Sampler (PAMS), that combines mutation steps with crossover steps. At each iteration, the algorithm chooses randomly between mutation and crossover steps. A mutation step is a standard Metropolis-Hastings update of a solution vector where the proposal is a random perturbation of the current solution (i.e., a small amount of typically Gaussian noise is added). The crossover steps create new proposals, which are then also accepted or rejected via Metropolis-Hastings. They propose both uniform crossovers and a variant of the snooker crossover, moving along the line connecting two current solutions. The selection and resampling step does not have an explicit counterpart in MCMC, but there are elements of selection involved in the MCMC versions of mutation and crossover, leading to similar net results.

A further step is Evolutionary Monte Carlo (EMC) (Liang and Wong, 2001), which combines MCMC, genetic algorithms, and parallel tempering. Just as with parallel tempering (from Section 14.3), a ladder of target functions is created,

$$f_i(x) = \frac{1}{Z(t_i)} \exp\{-Q(x)/t_i\},$$

where t serves as the temperature, $t_1 > \ldots > t_M$, Q is typically the negative of the unnormalized log-posterior (i.e., the log-likelihood plus the log-prior), and Z is the normalizing constant (which here is a function of t). The algorithm is a series of mutation, crossover, and exchange steps. Mutation, as before, is a standard additive noise Metropolis-Hastings update. Crossovers can be any of the standard varieties, such as k-point, uniform, or snooker crossovers. The exchange step proposes swapping solutions at different temperature levels. Typically, a pair of adjacent temperatures, t_i and t_{i+1}, are chosen, solutions x_i and x_{i+1} are swapped, and then this proposal is accepted according to the usual Metropolis-Hastings probability calculation. Multiple pairs can be attempted to be swapped either simultaneously or stepwise during the exchange step. The exchange step is the same as the moves between scales in parallel tempering and is analogous to the Metropolis-coupled swaps of the previous chapter. By combining parallel tempering and genetic algorithms, EMC is a method for exploring posteriors that has improved chances of escaping from local modes.

15.3.2 Multiresolution Versions

Building on the ideas of Evolutionary Monte Carlo and Metropolis coupling, Holloman et al. (2006) introduced an MCMC algorithm with multiscale genetic algorithm updates. A set of M Markov chains are run in parallel, nominally independently. M could be chosen to be equal to I, the number of scales, or as with the multiscale genetic algorithm of Section 15.2, one could ensure that at least two chains are started at each of the I scales. As with Metropolis coupling, information is shared across scales via MCMC updates, in this case crossover moves inspired by genetic algorithms.

Denote the chains by $\Gamma_1, \ldots, \Gamma_M$, so that for any $m \in \{1, \ldots, M\}$, Γ_m explores one part of the posterior: $\pi^{(i)}(\psi^{(i)} \mid \mathbf{y}^{(i)})$ for some $i \in \{1, \ldots, I\}$. The values of the random variables in any chain at a given time step determine the current state of that chain. We denote the state of the lth chain at time k as $\Gamma_l(\phi^l_{(k)}, \lambda^l_{(k)})$, where, as before, $\psi = (\phi, \lambda)$. ϕ represents the vector of parameters with a common interpretation across all scales, and λ contains all of the scale-specific parameters. It is unfortunate that so many superscripts and subscripts are necessary, separate indices are needed to distinguish (i) which of the M solutions it is, (ii) which scale the solution is on (with the accompanying data, likelihood, and prior), (iii) which MCMC iteration (time step) t it is at, and (iv) elements within the parameter vector (e.g., the first element may be a precision, the next a mean). Hence superscripts without parentheses specify which of the M solutions is under consideration, superscripts in parentheses specify the scale, subscripts in parentheses specify the iteration, and subscripts without parentheses specify the elements of a solution. We will try to use only one superscript and one subscript at a time, with the other values being implied by the context and usually not being modified. With this notation defined, a diagram of a portion of the algorithm for $M = 3$ is given in Figure 15.1.

$$\Gamma_1(\phi^1_{(1)}, \lambda^1_{(1)}) \xrightarrow{\text{MCMC}} \Gamma_1(\phi^1_{(2)}, \lambda^1_{(2)}) \qquad \Gamma_1(\phi^{1\prime}_{(3)}, \lambda^{1\prime}_{(3)}) \xrightarrow{\text{MCMC}} \Gamma_1(\phi^{1\prime}_{(4)}, \lambda^{1\prime}_{(4)}) \ldots$$
$$\Gamma_2(\phi^2_{(1)}, \lambda^2_{(1)}) \xrightarrow{\text{MCMC}} \Gamma_2(\phi^2_{(2)}, \lambda^2_{(2)}) \xrightarrow{\text{SWAP}} \Gamma_2(\phi^{2\prime}_{(3)}, \lambda^{2\prime}_{(3)}) \xrightarrow{\text{MCMC}} \Gamma_2(\phi^{2\prime}_{(4)}, \lambda^{2\prime}_{(4)}) \ldots$$
$$\Gamma_3(\phi^3_{(1)}, \lambda^3_{(1)}) \xrightarrow{\text{MCMC}} \Gamma_3(\phi^3_{(2)}, \lambda^3_{(2)}) \qquad \Gamma_3(\phi^3_{(3)}, \lambda^3_{(3)}) \xrightarrow{\text{MCMC}} \Gamma_3(\phi^3_{(4)}, \lambda^3_{(4)}) \ldots$$

Fig. 15.1. Example of a multiscale crossover within MCMC, where chains 1 and 2 attempt a swap in step 2.

As with the other algorithms that combine MCMC and genetic algorithms, this algorithm is a mix of standard Metropolis-Hastings MCMC updates (mutations) and swapping steps (crossovers). Here each chain advances using ordinary within-chain MCMC steps (mutations) for a fixed number of iterations until a swap step is reached. At that point, one or more pairs of chains attempt to swap values. In the diagram, a prime mark after a variable realization (e.g., $\lambda^{1\prime}_{(3)}$) indicates that the realization will take different values depending on whether the swap is successful or not. If the swap is unsuccessful, the primed variables take the same values as their unprimed counterparts at the previous time step (e.g., $\phi^{1\prime}_{(3)} = \phi^1_{(2)}$). If the swap is successful, the values of the primed variables take on the values proposed in the swap (e.g., $\phi^{1\prime}_{(3)} = \phi^2_{(2)}$ and $\phi^{2\prime}_{(3)} = \phi^1_{(2)}$). Note that since there are an odd number of chains in this example, the third chain does not take part in the swap. The next section describes the crossover step in more detail.

Multiscale Crossover Step

In a crossover step, two chains exchange part or all of the relevant parts of their solution vectors (the ϕ part). For now, we assume the chains are chosen with equal probability. We will later return to the possibility of using additional selection techniques in choosing the chains.

After pairing, proposals are formed from a crossover of the ϕ vectors of the two paired chains. Any of the previously described crossover moves can be used. One that we have found particularly effective is uniform crossovers with the probability of each element being swapped at 50%. The proposed values for a chain are marked with stars (e.g., ϕ^{1*} is a proposed ϕ vector for chain Γ_1). As with the Metropolis-coupled algorithm, acceptance is usually improved by proposing new values $\boldsymbol{\lambda}^*$ from some distributions $q^{(i)}(\boldsymbol{\lambda}^{j*} \mid \phi^{j*}, \mathbf{y}^{(i)})$ which are based on the proposed ϕ values for each chain j. The complete conditional distribution is an obvious choice when it is of a form that can be easily sampled. The proposal distributions $q^{(i)}$ may be different at each scale. For the swap depicted in Figure 15.1 where the first two chains attempt a swap, we accept or reject the swap according to the Metropolis-Hastings rule with probability

$$\frac{\pi^{(1)}(\phi^{1*}, \boldsymbol{\lambda}^{1*} \mid \mathbf{y}^{(1)})\pi^{(2)}(\phi^{2*}, \boldsymbol{\lambda}^{2*} \mid \mathbf{y}^{(2)})q^{(1)}(\boldsymbol{\lambda}^1 \mid \phi^1, \mathbf{y}^{(1)})q^{(2)}(\boldsymbol{\lambda}^2 \mid \phi^2, \mathbf{y}^{(2)})}{\pi^{(1)}(\phi^1, \boldsymbol{\lambda}^1 \mid \mathbf{y}^{(1)})\pi^{(2)}(\phi^2, \boldsymbol{\lambda}^2 \mid \mathbf{y}^{(2)})q^{(1)}(\boldsymbol{\lambda}^{1*} \mid \phi^{1*}, \mathbf{y}^{(1)})q^{(2)}(\boldsymbol{\lambda}^{2*} \mid \phi^{2*}, \mathbf{y}^{(2)})}$$

(or probability one if this ratio is larger than one). Holloman (2002) shows that this acceptance probability guarantees detailed balance in the MCMC chain.

We refer to moves of the form above as *crossover* swaps. In some cases, it can be useful to use swap steps that trade all elements. One could think of such a step as a uniform crossover with element swap probability of 100%. In this context, we refer to these steps as *full* swaps. In the context of the previous chapter, a full swap is equivalent to the Metropolis-coupled swap. These moves are also analogous to the exchange steps of Evolutionary Monte Carlo (where entire solutions were traded between temperatures). Full swap proposals can lead to greater exploration of the parameter space but are also more likely to be rejected (because more elements are being exchanged), so there is a trade-off between the types of swaps. In some of our implementations, when a swap step is called for, we randomly select between a full and a crossover swap.

One detail that can be expanded is the selection of the chains to be paired together. Earlier we assumed that they were picked with equal probability. One modification would be to require that pairs can only swap with adjacent scales, which typically improves acceptance probabilities but at the expense of lesser gains in mixing (i.e., allowing nonadjacent swaps can lead to additional improvement in mixing, but these swaps are less likely to be accepted). Another alternative plan involves appealing to the idea of selection from genetic algorithms and attempting to encourage the pairing of the most promising solutions. Instead of uniform probabilities, the two chains are chosen with probabilities proportional to their fitness measures. One possible fitness criterion would be the unnormalized posterior density of a realization on its portion of the full posterior. This criterion is valid if priors, likelihoods, and known values are similar on all scales, since the unknown normalizing constants for each scale should be about the same. If the likelihoods are not comparable across scales, then one could select chains in a two-step process: first, select a scale randomly with probabilities proportional to the number of available (unpaired) chains at that scale; second, select chains from that scale with probabilities proportional to their unnormalized likelihoods. Caution should be used when performing the selection with different fitness functions for different chains since all portions of the posterior are only known up to a normalizing constant. When in doubt, we recommend weighting chains equally. Note that if nonuniform probabilities are used to select the pairings of the chains, then a term for these selection probabilities must also be included in the Metropolis-Hastings acceptance ratio.

15.3.3 Implementational Concerns

Before moving on to an example, we first want to address a couple of implementational items. Depending on the setting, the user of these algorithms may need to make a choice of possible scales. The user will also need to choose an appropriate number of chains/solutions. Finally, we give a brief discussion of parallel computing.

In a truly multiscale problem, the scales will be determined by the data. In other problems, one could picture additional possible unobserved scales which could be used for interpretation or for computational purposes. This setting could also include the case where data are only observed on a single scale. When the user has the latitude to choose scales, the choice of both how many and which scales must be made. The simulated tempering literature provides some guidance for choosing the number of scales by analogy with their recommendations for choosing heated distributions. In simulated tempering schemes, it has been recommended that the temperatures be chosen such that the acceptance probability of moves between adjacent scales is between about 20% and 40% (Geyer and Thompson, 1995). This recommendation can be used as a guideline for selecting the scales to be used in modeling multiscale problems. This now leaves the choice of the endpoints, the finest and coarsest scales. The finest scale is the smallest resolution of interest to the practitioner. Typically, this is the finest resolution that is actually observed. The coarsest scale is usually chosen to be one where the Markov chain mixes easily, so that if MCMC was run only at that scale, it would quickly explore the whole space, not becoming unduly stuck in local modes.

Another consideration in choosing the scales is that of computational resources. Since at least one Markov chain is run at each scale, using more scales translates into additional computational effort. If the chains are run on a single machine, using too many scales can take prohibitively long. If the chains are run in parallel on separate processors, the user will quickly run up against the limit of the physical number of processors available. So there is a compromise between efficiency of moving between scales and efficiency in the allocation of available computational resources.

A similar trade-off needs to be made in choosing the number of chains (or, equivalently, the number of solution vectors). At the least, one chain is needed for each scale. But efficiency can often be improved by using more than one chain per scale. For a genetic algorithm, if there are too few chains, the algorithm may not have enough diversity in the population of chains to allow rapid exploration. At the other end of the spectrum, having too many chains wastes computational effort. For standard genetic algorithms, the usual recommendation is to set $M = 2 \times R$, where R is the length of the parameter vector. For multiscale MCMC implementations, premature convergence as found in genetic algorithms with too few solution vectors is not a large threat, so fewer chains can be used. Two chains per scale is a reasonable compromise.

This multiscale MCMC algorithm is easily adapted for parallel computing environments. Independent execution of mutation steps and fitness function evaluations is possible for each element of the population. Factorization of the posterior distribution into conditionally independent components allows simultaneous sampling of each of those components. Synchronization is only important for the swap steps, and not much data need to be passed at those steps. In addition, multiple swaps can be performed simultaneously in the MCMC algorithm. When pairing in parallel implementations, making as many nonoverlapping pairs as possible can be more efficient, but making several swaps simultaneously can make the acceptance probability very small if chains are chosen with probabilities proportional to some measure of fitness. In this situation, all proposed swaps must be accepted or rejected together. In contrast, if chains are chosen with equal probability, individual swap proposals between pairs of chains may be evaluated for acceptance or rejection independently, so performing several swaps simultaneously usually improves performance. Further details on performing several swaps simultaneously can be found in Holloman (2002).

Because there may be different amounts of data and different likelihoods for each scale, synchronization disparities can easily arise, leading to idle processors. It can pay to try to time the length of the run for each node so that they all work for approximately the same amount of time before a swap is considered. For instance, a scale with half the resolution of the next finer one could do two cycles of within-scale parameter updates (mutation and selection for the multiscale genetic algorithm or standard MCMC updates in the multiscale MCMC algorithm) in the time the finer scale does one. Additional information on issues in parallel MCMC can be found in Rosenthal (2000) and the references therein.

15.4 Example

Here we continue the example from Section 14.2.1 of fitting the first three years of the Fraser River data using a convolution model. This is not the best example, but we use it here for the sake of continuity. Since the model is fairly simple, there isn't much additional benefit from the crossover swaps compared with the full swaps of the previous chapter. The fitted values look quite similar to those in Figure 14.1 in the previous chapter. Also, there is typically not much additional reduction in autocorrelation, as shown below in Figure 15.2. In this case, the methods of the previous chapter have already taken advantage of mixing across scales, and there is little left to be gained by also using crossover swaps. However, for more complicated problems there can be significant gains.

An additional example is the single photon emission computed tomography (SPECT) example in Chapter 17. Of interest is a pixelated map of image intensities, which must be reconstructed from photon counts collected by a rotating gamma camera. A Markov random field prior is used for the intensity map, and a physical forward model links the hypothesized intensities to the observed photon counts. Fitting of the latent intensity map pixels is done with MCMC, but mixing can be poor because of the high correlation between the unknown parameters. As with many imaging problems, it is expected that neighboring pixels will be highly correlated, as otherwise the image would just be a random collection of numbers, not a coherent image. These correlations make it difficult to create good Metropolis-Hastings proposals, as the proposal must make enough changes in order to be different enough from the current values that it makes progress in exploring the space, but not too different from the current values so that it is rejected. In this case, too "different" is often achieved because neighboring pixels are not kept sufficiently similar in a proposal. The multiscale methods of this chapter and the previous chapter provide an efficient mechanism for creating good proposals. These proposals can retain enough of the spatial qualities to allow reasonable acceptance probabilities, yet provide enough movement so that the posterior can be explored with a practical amount of computational effort. The underlying imaging problem here is not inherently multiscale, but by embedding the problem in a multiscale framework, the inference can be done more efficiently through multiscale MCMC. Chapter 17 provides additional details.

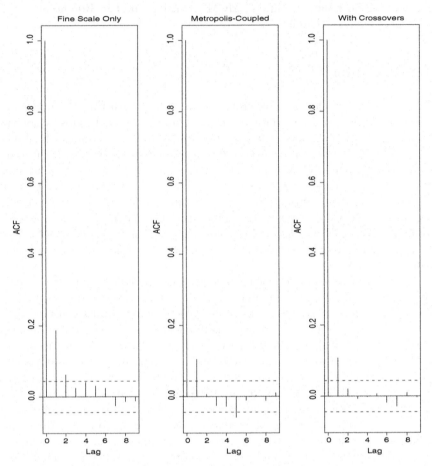

Fig. 15.2. Comparison of autocorrelation plots for a latent value.

Part V

Case Studies

Soil Permeability Estimation

16.1 Introduction

Many important applications in the cleanup of contaminated soil and in the production of petroleum from oil fields require the understanding of soil structure and prediction of fluid flow patterns. Subsurface flow is dependent on a number of factors, with one of the primary ones being the permeability field. Permeability is a measure of how easily fluid moves through the porous medium at a point, and it varies spatially. In this chapter, we treat permeability as a constant at a point, but in more detailed applications it may be tensor-valued to account for the fact that soil structure can be directional and so fluid may flow more easily in some directions than in others.

Permeability estimation is important both for its own sake and for helping to determine flow patterns. In environmental cleanup applications, permeability estimates are critical for two parts: for identifying the likely location and distribution of the contaminant and for designing the remediation operation (James et al., 1997; Jin et al., 1995). The contamination can be treated without having to dig up the ground by pumping water containing a surfactant through the affected area. The water can then be treated to remove the contaminants, a much quicker and less costly operation than excavating and treating the soil itself. However, effective use of this procedure requires good knowledge of the permeability field. In petroleum production, primary production occurs when a well is drilled into an underground oil field and the oil rises to the surface under its own pressure. Typically, much less than half of the oil can be extracted in this manner. Secondary and tertiary recovery efforts require increasing the pressure within the oil field to cause more oil to rise to the surface. This increase in pressure is achieved by pumping either water or natural gas (a frequent by-product of oil production) into the field. Effective recovery procedures rely on good permeability estimates, as one must be able to identify high-permeability channels and low-permeability barriers (Xue and Datta-Gupta, 1996).

There are in general two types of available information for the estimation of the permeability field: static data and dynamic data. The static information about the permeability field comes from a variety of sources, such as geological studies, well tests, and core samples. Each of these sources provides information at different levels of aggregation or resolution: geological studies (for example, seismic studies) provide information at a very coarse level of resolution. Well tests provide information at an intermediate level, giving more information in the local area around a well but also some information farther from the well. Core logs, either taken to a laboratory for direct permeability measurements or measuring resistivity in the bore hole, give information at a very fine level. These core measurements only provide information at the sampling point, so all spatial information must be inferred from the model. It is also worth pointing out that while laboratory measurements are the only direct estimates of permeability, they are also expensive, variable, and destructive. Drilling a core sample, returning it to a lab, and analyzing it is a labor-intensive process. There can be large measurement errors induced because the soil in a core can act differently than it did in the ground—drilling the core is disruptive to the soil one is trying to measure. And the core is destructive to the remaining field in that now there is a hole in the ground and often the surrounding material will collapse into the hole. The permeability and the water flows can be radically altered by the very process of removing the core. Thus there is a trade-off between level of resolution and coverage of the field. Higher-resolution data may be available for restricted parts of the field, while the coarse-resolution data are generally available for the whole field. Here, we assume that the static data are either direct measurements or known deterministic and invertible transformations of measurements of the permeability field at the corresponding scales of resolution.

The dynamic data are obtained from results of tracer experiments performed in an aquifer or from the history of oil production of a petroleum field. As opposed to the simple relationship with the static data, the permeability field affects the fluid flow and therefore the dynamic data in a highly nonlinear way. If the permeability field and other physical characteristics of the porous media were known, physical models such as Darcy's Law, Fick's Law, and the law of conservation of mass could be used to solve the forward problem; that is, to predict the fluid flow and the dynamic data (Gelhar, 1993). These physical models are described by a system of differential equations that can be solved numerically by computer codes called fluid flow simulators (FFS) (King and Datta-Gupta, 1998; Vasco and Datta-Gupta, 1999). In this chapter, relatively fast simulators are used to compute approximations for the expected dynamic data for permeability fields partitioned into discrete grid-blocks. An extremely important point for practical implementations is that as the resolution of the partition increases, the FFS computes more accurate solutions but becomes slower.

The use of dynamic data for the estimation of the permeability field is known in the subsurface hydrology literature as the inverse problem. This terminology arises in contrast to the forward problem, which is the solution of the flow equations when the permeability is known, as discussed in the previous paragraph. The inverse problem is the opposite — having observed dynamic data, one tries to infer the permeability field. To do so requires iterative use of the solution to the forward problem in that one guesses a value for the permeability field, solves the forward problem to see what flows would result, and then goes back and adjusts the initial guess at the permeability field, iterating until the predicted flows match the observed data closely enough. References on the subsurface hydrology inverse problem include Neuman and Yakowitz (1979), Yeh (1986), Oliver (1994), Craig et al. (1996), Hegstad and Omre (1997), Oliver et al. (1997), Vasco and Datta-Gupta (1999), Floris et al. (2001), and Lee et al. (2002). The inverse problem is ill-posed; i.e., there are an infinite number of permeability field solutions whose expected dynamic data exactly match the observed dynamic data. Many of those solutions are too rough to be plausible, and proposed approaches generally impose some stochastic regularity constraints on the permeability field. For example, Bonet-Cunha et al. (1998) and Oliver et al. (1997) assume that the permeability field follows a Gaussian process model, while Lee et al. (2002) consider allowing the permeabilities to follow a Markov random field process. While the Gaussian process model can imply permeability fields that are too smooth, the Markov random field assumption can lead to an undesirably fast decay of the spatial correlation. Thus we turn to multiscale approaches.

Here we present two different approaches. First, the multiscale models of Chapter 10 are used as priors for permeability fields. These multiscale models have the local behavior of Markov random fields but globally emulate long memory processes, being well suited to capture global features of the permeability field. In combination with this multiscale prior, corresponding likelihood functions for the high-dimensional random field parameters representing the permeability field at each level or resolution are computed with the help of the FFS embedded in an MCMC scheme. This statistical framework uses the faster solutions at the coarser levels to guide the solutions at the finer levels. Results in the next section appeared originally in Ferreira et al. (2003) and Ferreira et al. (2005). Section 16.2 presents the multiscale permeability model. Section 16.2.1 describes the propagation of static information through the different levels of resolution, and Section 16.2.2 discusses the incorporation of the dynamic data. A one-dimensional permeability field example appears in Section 16.2.3, and Section 16.2.4 presents the application of the multiscale framework to several two-dimensional permeability fields.

The second approach, presented in Section 16.3, is that of the implicit multiscale models of Chapter 14. The setup of the problem is analogous to that in Section 16.2.4. Rather than making modeling assumptions about the links between the scales, we can link them through the model fitting, achieving good fits and improved computational efficiency. Some of these results originally appeared in Higdon et al. (2002).

16.2 Multiscale Modeling

In order to estimate a true continuous permeability field, the field is discretized in coarse, intermediate, and fine versions denoted respectively by $\mathbf{x}_0$, $\mathbf{x}_1$, and $\mathbf{x}_2$. The multiscale random field of Chapter 10 is used as a prior for the multiscale permeability field. More specifically, the prior of the multiscale permeability field is of the form

$$p(\mathbf{x}_0|\mu, \tau_0, \alpha_0)p(\mathbf{x}_1|\mathbf{x}_0, \mu, \tau_1, \alpha_1, \delta_1)p(\mathbf{x}_2|\mathbf{x}_1, \mu, \tau_2, \alpha_2, \delta_2),$$

where

$$p(\mathbf{x}_0|\mu, \tau_0, \alpha_0) = N(\mathbf{x}_0|\mu\mathbf{1}_{n_0}, \mathbf{Q}_0),$$

$$p(\mathbf{x}_1|\mathbf{x}_0, \mu, \tau_1, \alpha_1, \delta_1) = N(\mathbf{x}_1|\mu\mathbf{1}_{n_1} + \mathbf{B}_1(\mathbf{x}_0 - \mu\mathbf{1}_{n_0}), \boldsymbol{\Sigma}_1 - \mathbf{B}_1\mathbf{W}_1\mathbf{B}_1'),$$

and

$$p(\mathbf{x}_2|\mathbf{x}_1, \mu, \tau_2, \alpha_2, \delta_2) = N(\mathbf{x}_2|\mu\mathbf{1}_{n_2} + \mathbf{B}_2(\mathbf{x}_1 - \mu\mathbf{1}_{n_1}), \boldsymbol{\Sigma}_2 - \mathbf{B}_2\mathbf{W}_2\mathbf{B}_2'),$$

where μ is the mean level of the field, $\mathbf{B}_1$, $\mathbf{B}_2$, $\mathbf{W}_1$, and $\mathbf{W}_2$ are as defined in Theorem 10.2, and $\mathbf{Q}_0$, $\boldsymbol{\Sigma}_1$, and $\boldsymbol{\Sigma}_2$ are covariance matrices of proper Markov random fields as defined by Equations (10.1) and (10.2).

Denote the available information by

- $\mathbf{p}_{obs}$ = observed dynamic data;
- $\mathbf{d}_0$ = measurement of permeability at the coarse scale;
- $\mathbf{d}_1$ = measurement of permeability at the intermediate scale;
- $\mathbf{d}_2$ = measurement of permeability at the fine scale.

In addition, it is assumed that

- $\mathbf{p}_{obs}|\mathbf{x}_2 \sim N(f(\mathbf{x}_2), \sigma_\varepsilon^2\mathbf{I})$,
- $\mathbf{d}_0|\mathbf{x}_0 \sim N(\mathbf{x}_0, \mathbf{S}_0)$,
- $\mathbf{d}_1|\mathbf{x}_1 \sim N(\mathbf{x}_1, \mathbf{S}_1)$, and
- $\mathbf{d}_2|\mathbf{x}_2 \sim N(\mathbf{x}_2, \mathbf{S}_2)$,

where $\mathbf{S}_0$, $\mathbf{S}_1$, and $\mathbf{S}_2$ are the known covariance matrices of the measurement errors at the respective levels of resolution and $f(\mathbf{x})$ is the expected dynamic data for the permeability field $\mathbf{x}$. Here it is assumed that the expected dynamic data can be computed by the FFS with negligible approximation error. Typically, $\mathbf{S}_0$, $\mathbf{S}_1$, and $\mathbf{S}_2$ are diagonal matrices or covariance matrices obtained from a geostatistical analysis of the static data.

As a result, the posterior density will be proportional to

$$p(\mathbf{p}_{obs}|\mathbf{x}_2, \sigma_\varepsilon^2)p(\mathbf{d}_2|\mathbf{x}_2)p(\mathbf{d}_1|\mathbf{x}_1)p(\mathbf{d}_0|\mathbf{x}_0)p(\mathbf{x}_0|\mu, \tau_0, \alpha_0)$$

$$p(\mathbf{x}_1|\mathbf{x}_0, \mu, \tau_1, \alpha_1, \delta_1)p(\mathbf{x}_2|\mathbf{x}_1, \mu, \tau_2, \alpha_2, \delta_2)$$

$$p(\mu)p(\tau_0, \alpha_0)p(\tau_1, \alpha_1, \delta_1)p(\tau_2, \alpha_2, \delta_2).$$

This posterior density is explored with an MCMC scheme with each iteration divided in two main steps. In the first step, all the information about the static data $\mathbf{d}_0$, $\mathbf{d}_1$, and $\mathbf{d}_2$ is propagated from the finer to the coarser levels by analytically integrating out the finer levels. In the second step, the dynamic data are incorporated in the analysis in a cascade way from coarser to finer levels of resolution. First, the dynamic data are used to generate the coarser levels by using the FFS running at those levels. After that, the finer levels are generated conditional on the generated coarser levels. The incorporation of the dynamic data from coarser to finer levels brings two advantages. The first advantage is that the running time of the FFS at the coarser levels is faster than at the finer levels. The second advantage is that the MCMC for the coarser levels converges much faster than for the finer levels because the marginal posterior distribution of coarser levels has fewer local maxima than those of finer levels. As a result, the results at coarser levels guide the simulation at finer levels, and the main consequence is faster convergence at finer levels as compared with convergence of traditional MRF schemes.

Thus, the main idea of this MCMC algorithm is the following: the coarse level $\mathbf{x}_0$ is simulated from $p(\mathbf{x}_0|\mathbf{p}_{obs}, \mathbf{d}_0, \mathbf{d}_1, \mathbf{d}_2)$; that is, $\mathbf{x}_1$ and $\mathbf{x}_2$ are integrated out analytically before the simulation of $\mathbf{x}_0$. Then, $\mathbf{x}_1$ is simulated from $p(\mathbf{x}_1|\mathbf{x}_0, \mathbf{p}_{obs}, \mathbf{d}_0, \mathbf{d}_1, \mathbf{d}_2)$ with $\mathbf{x}_2$ integrated out. Finally, $\mathbf{x}_2$ is simulated from its full conditional $p(\mathbf{x}_2|\mathbf{x}_1, \mathbf{x}_0, \mathbf{p}_{obs}, \mathbf{d}_0, \mathbf{d}_1, \mathbf{d}_2)$. The next sections present the expressions of these distributions.

16.2.1 Static Information Propagation

This section describes how to send the measurement information from finer to coarser levels. As the generalization to more levels is straightforward, here only two levels are considered. For notational simplicity, we omit the dependence of the density functions on the hyperparameters. Thus, the objective is to obtain $p(\mathbf{x}_0|\mathbf{d}_0, \mathbf{d}_1)$. Hence

$$p(\mathbf{x}_0|\mathbf{d}_0, \mathbf{d}_1) \propto p(\mathbf{d}_0, \mathbf{d}_1|\mathbf{x}_0)p(\mathbf{x}_0)$$

$$= p(\mathbf{d}_0|\mathbf{x}_0)p(\mathbf{x}_0) \int p(\mathbf{d}_1|\mathbf{x}_1)p(\mathbf{x}_1|\mathbf{x}_0)d\mathbf{x}_1$$

$$\propto \frac{N(\mathbf{d}_0|\mathbf{x}_0, \mathbf{S}_0)N(\mathbf{x}_0|\mu\mathbf{1}_{n_0}, \mathbf{Q}_0)}{N(\mathbf{x}_0|\mu\mathbf{1}_{n_0}, \mathbf{A}_1\boldsymbol{\Sigma}_1\mathbf{A}_1' + \delta_1\mathbf{I}_{n_0})}N(\mathbf{x}_0|\mathbf{b}_0, \mathbf{B}_0),$$

where $\mathbf{A}_1$ is the matrix that performs the coarsening operation as in Equation (10.3),

$$\mathbf{B}_0 = \mathbf{A}_1(\mathbf{S}_1^{-1} + \boldsymbol{\Sigma}_1^{-1})^{-1}\mathbf{A}_1' + \delta_1\mathbf{I}_{n_0},$$

and

$$\mathbf{b}_0 = \mathbf{A}_1(\mathbf{S}_1^{-1} + \boldsymbol{\Sigma}_1^{-1})^{-1}(\mathbf{S}_1^{-1}\mathbf{d}_1 + \tau_1\alpha_1\mu\mathbf{1}_{n_1})$$

since $\boldsymbol{\Sigma}_1^{-1}\mathbf{1}_{n_1} = \tau_1\alpha_1\mathbf{I}_{n_1}$.

16.2.2 Dynamic Data Incorporation

After propagating the measurement data from finer to coarser levels, the algorithm incorporates the dynamic data from coarser to finer levels. The first step is the simulation of $\mathbf{x}_0$, conditional on the pressure and the measurement data. It is assumed that, conditional on the coarse level, the dynamic data $\mathbf{p}_{obs}$ are independent of the measurements at the fine and intermediate levels. This assumption allows use of the FFS at different levels of resolution for the simulation of each level. Under this assumption, the conditional distribution of $\mathbf{x}_0$ given $\mathbf{p}_{obs}$, $\mathbf{d}_0$, $\mathbf{d}_1$, and $\mathbf{d}_2$ is

$$p(\mathbf{x}_0|\mathbf{p}_{obs}, \mathbf{d}_0, \mathbf{d}_1, \mathbf{d}_2) \propto p(\mathbf{p}_{obs}, \mathbf{d}_0, \mathbf{d}_1, \mathbf{d}_2|\mathbf{x}_0)p(\mathbf{x}_0)$$
$$\propto p(\mathbf{p}_{obs}|\mathbf{x}_0)p(\mathbf{x}_0|\mathbf{d}_0, \mathbf{d}_1, \mathbf{d}_2).$$

Note that $p(\mathbf{x}_0|\mathbf{d}_0, \mathbf{d}_1, \mathbf{d}_2)$ can be easily obtained using the result outlined in Section 16.2.1. Moreover, running the FFS at the coarse level provides a good approximation for $p(\mathbf{p}_{obs}|\mathbf{x}_0)$, that is, $p(\mathbf{p}_{obs}|\mathbf{x}_0) \approx N(\mathbf{p}_{obs}|f(\mathbf{x}_0), \sigma_{\varepsilon 0}^2 \mathbf{I})$. This avoids the cumbersome computation of the integral in $p(\mathbf{p}_{obs}|\mathbf{x}_0) = \int \int p(\mathbf{p}_{obs}|\mathbf{x}_2) \, p(\mathbf{x}_2|\mathbf{x}_1)p(\mathbf{x}_1|\mathbf{x}_0)d\mathbf{x}_1 d\mathbf{x}_2$, which is very complicated due to the nonlinear relationship between $\mathbf{p}_{obs}$ and $\mathbf{x}_2$.

Moreover, assuming $\mathbf{p}_{obs}$ independent of $\mathbf{d}_2$ given $\mathbf{x}_1$ implies

$$p(\mathbf{x}_1|\mathbf{x}_0, \mathbf{p}_{obs}, \mathbf{d}_0, \mathbf{d}_1, \mathbf{d}_2) \propto p(\mathbf{p}_{obs}|\mathbf{x}_1)p(\mathbf{x}_1|\mathbf{x}_0, \mathbf{d}_0, \mathbf{d}_1, \mathbf{d}_2)$$

$$\propto p(\mathbf{p}_{obs}|\mathbf{x}_1)p(\mathbf{x}_1|\mathbf{x}_0)p(\mathbf{d}_1|\mathbf{x}_1) \int p(\mathbf{d}_2|\mathbf{x}_2)p(\mathbf{x}_2|\mathbf{x}_1)d\mathbf{x}_2.$$

Again, the result of Section 16.2.1 is used to compute $\int p(\mathbf{d}_2|\mathbf{x}_2)p(\mathbf{x}_2|\mathbf{x}_1)d\mathbf{x}_2$, bringing the measurement information about the finest level to the intermediate level. In addition, $p(\mathbf{p}_{obs}|\mathbf{x}_1)$ can be approximated by running the FFS at the intermediate level and assuming that $p(\mathbf{p}_{obs}|\mathbf{x}_1) \approx N(\mathbf{p}_{obs}|f(\mathbf{x}_1), \sigma_{\varepsilon 1}^2 \mathbf{I})$.

Finally, the conditional distribution of $\mathbf{x}_2$ given $\mathbf{x}_1$, $\mathbf{x}_0$, $\mathbf{p}_{obs}$, $\mathbf{d}_0$, $\mathbf{d}_1$, and $\mathbf{d}_2$ is

$$p(\mathbf{x}_2|\mathbf{x}_1, \mathbf{x}_0, \mathbf{p}_{obs}, \mathbf{d}_0, \mathbf{d}_1, \mathbf{d}_2) \propto p(\mathbf{p}_{obs}|\mathbf{x}_2)p(\mathbf{x}_2|\mathbf{x}_1)p(\mathbf{d}_2|\mathbf{x}_2).$$

16.2.3 One-Dimensional Permeability Estimation

This section presents an application of the multiscale framework to the estimation of a synthetic 1-D permeability field. Some real-world problems can be cast in this one-dimensional framework. One such example is the estimation of the permeability field of incised valley oil fields, that is, petroleum fields that developed in buried ancient river beds. See Stephen and Dalrymple (2002) for an example of an incised valley oil field.

Porous media

Production
well

1280 m

Fig. 16.1. Design of the reservoir.

The design of the synthetic field is as follows. The length of the field is equal to 1280 meters, and there is only one production well located in the right end of the field, as depicted in Figure 16.1. There are three discretized versions of the field, x_0, x_1, and x_2, partitioning it into 8, 32, and 128 grid-blocks, respectively. Moreover, it is assumed that the fine discretized level x_2 is equal to the truth, and inference will be done only for the coarser resolutions x_0 and x_1.

In order to reduce the dimension of the problem of prior specification, we assume a priori independence of the parameters, allowing us to focus on the specification of the prior for each parameter individually.

Several aspects distinctive to the multiscale approach to the 1-D fluid flow problem facilitate the specification of informative priors for several parameters of interest. The most important of these characteristics is that the permeability field is modeled on the logarithm scale. Thus, reasoning on the relative variability between neighbor grid-blocks within and across different levels of resolution leads to straightforward specification of the priors for τ_0, τ_1, and δ_1. Considerations on the smoothness of the log-permeability field at each resolution level lead to the specification of the priors for the spatial dependence parameters α_0 and α_1. Physical considerations assist the specification of the prior for the overall log-permeability mean μ. The variances $\sigma_{\varepsilon 0}$ and $\sigma_{\varepsilon 1}$ are assigned vague priors. More specifically, the priors used in this particular application are $\tau_0 \sim Ga(3.0, 1.1)$, $\alpha_0 \sim Ga(10, 1)$, $\tau_1 \sim Ga(57, 0.14)$ $\alpha_1 \sim Ga(0.2, 20)$, $\mu \sim N(5.0, 1.21)$, $\sigma_{\varepsilon 0} \sim IG(5 \times 10^{-5}, 5 \times 10^{-5})$, $\sigma_{\varepsilon 1} \sim IG(5 \times 10^{-5}, 5 \times 10^{-5})$, and $\delta_1 \sim IG(57, 0.14)$.

Figure 16.2 presents the original synthetic permeability field with 512 grid-blocks and coarser versions with 64 and 8 grid-blocks, darker colors corresponding to lower values and lighter colors corresponding to higher values. The permeability of the ith grid-block of the synthetic field has a value of $\exp(-3.1710^{-5}i^2 + 0.01625i + 3.91)$, $i = 0, \ldots, 511$, $i = 1$ and $i = 512$ corresponding respectively to the left and right ends of the field. Other physical characteristics of the reservoir, such as number of phases, initial pressure,

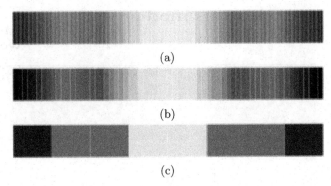

Fig. 16.2. Permeability field. (a) Original field (512 grid-blocks); (b) intermediate field (64 grid-blocks); (c) coarse field (8 grid-blocks). Lighter colors correspond to higher values.

porosity, production rate, cross-sectional area, and compressibility, are assumed constant and known. In addition, there is only one production well located in the right end of the field. The static data at the coarse level are $\mathbf{d}_0 = (4.4, 5.1, 5.6, 5.9, 6.0, 5.7, 5.2, 4.3)$ and $\mathbf{S}_0 = 0.1^2\mathbf{I}$. There are static data at the intermediate level only for the grid-blocks 1, 9, 17, 25, 33, 41, 49, 57, and 64 with variance equal to 0.004, with recorded values equal to 3.97, 4.86, 5.50, 5.88, 5.99, 5.85, 5.45, 4.78, and 3.99.

Depending on the production regime, the dynamic data can be either the flow rate or the pressure at the production well. In this example, the flow rate is kept constant and the dynamic data are given by the pressure curve at the production well. The FFS was used to compute the pressure curve for the first two days of production, and measurement errors with variance $\sigma_\varepsilon^2 = 1.0$ were added to the pressure curve, shown in Figure 16.3.

Figure 16.4 presents the coarse version of the original field and the posterior mean of the coarse level of resolution. Analogously, Figure 16.5 presents the intermediate version of the original field and the posterior mean of the intermediate level of resolution. At both resolution levels, the posterior mean is very close to the original permeability field, the main difference being that the estimation at the intermediate level of resolution recovers a higher degree of detail.

One of the main advantages of the Bayesian framework coupled with MCMC technology is the uncertainty characterization. The uncertainty present in the inference about the permeability field can be summarized by the posterior variance of each grid-block. Figure 16.6 presents the posterior variance across the field at the coarse and intermediate levels of resolution. At both levels of resolution, as the production well is located in the right end of the field, the incorporation of the dynamic data results in smaller posterior variances for grid-blocks closer to that region. In particular, the ratio between the

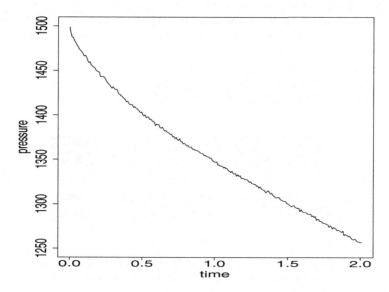

Fig. 16.3. Dynamic data for permeability estimation. Pressure curve through time.

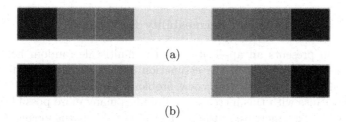

Fig. 16.4. Estimation at the coarse level of resolution. (a) Original coarse field; (b) posterior mean.

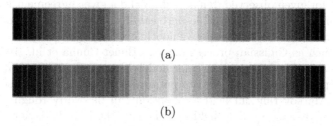

Fig. 16.5. Estimation at the intermediate level of resolution. (a) Original intermediate field; (b) posterior mean.

posterior variances of the coarse grid-blocks 1 (left end) and 8 (right end) is about 50, demonstrating the reduction in uncertainty due to the proximity of the production well. At the intermediate level, the existence of static data for the grid-blocks 1, 9, 17, 25, 33, 41, 49, 57, and 64 implies smaller posterior variances at these grid-blocks than at their neighbors. In addition, as the measurement information is equally spaced, the transfer of information along the field causes periodic behavior of the posterior variances.

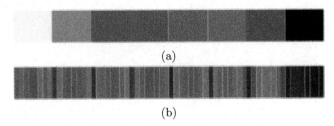

(a)

(b)

Fig. 16.6. Uncertainty in the estimation of the permeability field. Posterior variance across the field (a) at the coarse level and (b) at the intermediate level.

16.2.4 Two-Dimensional Permeability Estimation

This section presents an application of the multiscale random field models presented in Chapter 10 to the estimation of 2-D permeability fields. The increase in dimensionality poses new problems for the estimation of permeabilities because with this increase the fluid has many more possible paths to flow through. This increase in the number of possible paths means, for example, that a limited number of small areas of very low permeability have little effect on the fluid flow, as the fluid can go around those areas. As a consequence, the problem of permeability field estimation is severely ill-posed in the sense that for given observed dynamic data and a given agreement level, there are an infinite number of possible permeability fields. Thus one must make modeling assumptions that impose smoothness constraints on the permeability field, such as Gaussian process models (Bonet-Cunha et al., 1998; Oliver et al., 1997) or Markov random field models (Lee et al., 2002). However, the Gaussian process model can lead to permeability fields that are too smooth, while the Markov random field assumption can be overly rough. As an alternative, multiscale random field models have the local behavior of Markov random fields but globally emulate long memory processes. Thus, here we consider multiscale random field models as priors for permeability fields.

The setup of the examples in this section is the same as in Lee et al. (2002). In order to estimate the permeability field of an aquifer, the following experiment is performed. Water is pumped into the aquifer through one or more injection wells at a fixed rate. The water is extracted by producer wells while keeping constant the pressure at the bottom of each well. After equilibrium is reached, a tracer is injected at the injection well(s) and the concentration of the tracer is measured over time at each of the producer wells. Lee et al. (2002) found that the first time of arrival of the tracer at each well, called the breakthrough time, is practically a sufficient statistic for a unimodal concentration curve. In the petroleum engineering literature, Vasco et al. (1998) and Yoon et al. (1999) have also used breakthrough times as summaries of the concentration curves. Thus, here breakthrough times are used as the dynamic data.

More specifically, the setup in the first two examples is an inverted 9-spot pattern, that is, it is a square field with a single injection well in the center of the field and eight producer wells, one at each corner and one on the middle of each edge. Thus there are eight breakthrough times to estimate the whole permeability field. Denote by $\mathbf{p}_{obs}$ the dynamic data vector with the eight breakthrough times. These dynamic data are incorporated in the estimation of the permeability field at the different resolution levels using the results presented in Section 16.2.2. It is assumed that all other important physical quantities such as initial pressure and porosity are known. Moreover, it is assumed an incompressible medium and fluid, and an ideal tracer.

The permeability field affects the fluid flow and therefore the dynamic data in a highly nonlinear manner. When the physical characteristics of the porous media, such as the permeability field and porosity, are known, physical models such as Darcy's Law and the law of conservation of mass can be used to solve the forward problem. A fluid flow simulator (FFS) numerically solves these systems of differential equations. In the remainder of this chapter, we use the S3D streamtube code of King and Datta-Gupta (1998), which is fast enough to make Markov chain Monte Carlo solutions practical. Similar to Lee et al. (2002), a Gaussian likelihood for the logarithm of the breakthrough times is used, with each log-breakthrough time conditionally independent with mean equal to the value obtained from the FFS. The results of Section 16.2.2 lead to the use of the FFS at different levels of resolution in order to accelerate the procedure for estimation of the permeability field. More specifically, the algorithm runs first at the coarse level, and then at the fine level conditional on the coarse-level results. As a consequence, the algorithm converges faster at the fine level when compared with algorithms based on traditional Markov random field models, analogous to results on multigrid methods (see, for example, Briggs et al., 2000).

As there is no static data in this set of examples and as the dynamic data provide very little information about the smoothness of the field, the hyperparameters of the model have to be specified a priori on a case-by-case basis. Thus, in the following examples, the hyperparameters of the multiscale model are kept constant. As we model the log-breakthrough times, the variances σ_{ε_0} and σ_{ε_1} are related to the amount of allowed relative differences between the observed and the adjusted breakthrough times. Thus, it is reasonably easy to assign informative priors for σ_{ε_0} and σ_{ε_1}. In general, their simulated values are large in the beginning of the MCMC and become smaller as the multiscale random field realizations converge to a draw from their posterior distribution. This behavior of the traces of σ_{ε_0} and σ_{ε_1} leads to an effect similar to simulated annealing for the simulation of the different levels of the field, with higher temperatures in the beginning and cooler temperatures at the end of the chain.

The following sections present three examples of applications of the multiscale framework for the estimation of 2-D permeability fields. The first two examples are based on dynamic data simulated from synthetic linear and Gaussian log-permeability fields. These cases are informative because the true underlying field is known and so can be compared with the fitted values. The third example refers to data on breakthrough times obtained from a real experiment at the Hill Air Force Base in Utah.

Linear Field

Figure 16.7 presents a linear log-permeability field with lower permeabilities in the southwest corner of the field and higher permeabilities in the northeast corner, as well as some results of the multiscale analysis. In the plots, darker colors indicate higher permeability values.

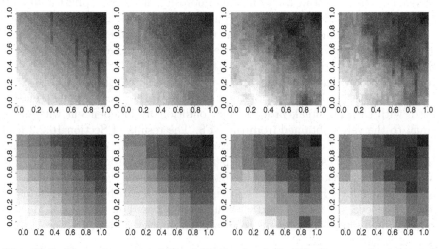

Fig. 16.7. Linear log-permeability field: fine level (first line), coarse level (second line), original field (first column), posterior mean (second column), and realizations (third and fourth columns).

The original 32×32 log-permeability field is shown in the upper left of Figure 16.7. The original field was used as input for the fluid flow simulator, generating 8 breakthrough times. Using the multiscale framework outline in this chapter, these 8 breakthrough times were used to check how well we can estimate the original field.

The lower left of Figure 16.7 shows the 8×8 coarse version of the original field. The second column shows the posterior means of the fine and coarse levels of the log-permeability field. As can be seen in the figure, the posterior mean recovers the original field very well.

The last two columns show some realizations of the fine and coarse levels of the permeability field. There is a reasonable amount of variability between the realizations. It is remarkable how the coarse-level realizations drive the fine-level realizations and how features in the coarse level are reflected in the fine level.

Gaussian Field

Figure 16.8 presents the logarithm of a Gaussian permeability field as well as some results of the multiscale analysis. In the plots, darker colors indicate higher permeability values. The layout in Figure 16.8 is the same as in Figure 16.7.

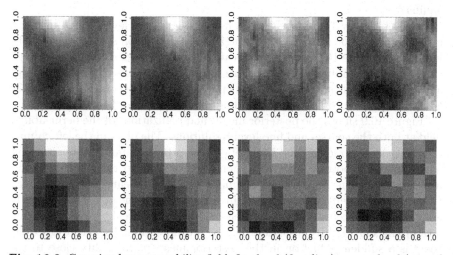

Fig. 16.8. Gaussian log-permeability field: fine level (first line), coarse level (second line), original field (first column), posterior mean (second column), and realizations (third and fourth columns).

The multiscale framework recovers the original field very well. The posterior mean has most of the major features of the original field, such as the regions of lower permeability in the southeast corner and in the center-north, and the region of higher permeability in the central region and in the center-south.

It is really amazing to see how the fine-level realizations derive from the coarse-level realizations and how the fine-level realizations are much smoother than their coarse counterparts. There is a lot of variability between the fine-level realizations, but they still show some of the main features of the original permeability field. Moreover, some artifacts such as very small regions of lower or higher permeability appear in each of the realizations. The appearance of these artifacts happens because the fluid can easily go around small regions of low permeability.

Hill Air Force Base Dataset

In this example, the multiscale framework is used to estimate the permeability field of a test site where the ground contains several contaminants at Hill Air Force Base in Utah. Multiple tracer experiments were run to estimate the physical characteristics in order to support the cleanup of the aquifer (Annable et al., 1998; Yoon, 2000). The focus here is on the data from an experiment with a conservative tracer whose interaction with the contaminants is negligible, so that the flow data reflects the permeability structure of the aquifer.

The 14 foot by 11 foot test site is modeled with two levels of resolution: a coarse resolution on a 14 by 11 grid and a fine resolution on a 28 by 22 grid. The left plots of Figure 16.9 contain the locations of the wells and the breakthrough times at each sampling well. There are four injection wells along a short edge and three production wells along the opposite edge. The concentration of the tracer is measured only at five sampling wells in the middle of the site.

Figure 16.9 shows the posterior means and realizations of the permeability field at both the coarse- and fine-level resolutions. The plots in the left column are the posterior means, and the plots in the other columns are realizations. The upper and bottom lines correspond respectively to the fine and coarse levels of resolution. As the lower sampling well has the earliest breakthrough time, the lower left region has the highest posterior mean permeability. In addition, as the central sampling well has the latest breakthrough time and a sampling well slightly to its left has the second earliest breakthrough time, there is a sudden reduction in the posterior mean permeability close to the central well. Moreover, the top left region has reasonably high posterior mean permeability and the right region has intermediate level posterior mean permeability. Note that even though the posterior means at both resolution levels are reasonably smooth, the posterior realizations are quite noisy, indicating that many different pathways are consistent with the observed breakthrough times. Nevertheless, the posterior realizations clearly indicate that the region close to the central well has the lowest permeability.

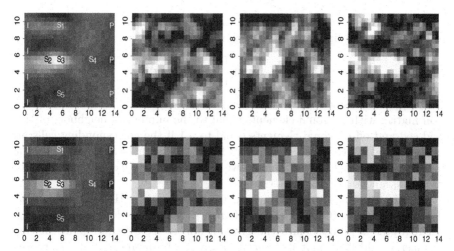

Fig. 16.9. Hill Air Force Base example. Log-permeability field: fine level (first line), coarse level (second line), posterior mean (first column), and realizations (second, third, and fourth columns).

16.3 Implicit Multiscale Methods

An alternative modeling scheme to that in the previous section is to use independent priors at each level and then to link them implicitly via the computation, as with the Metropolis-coupled methods from Chapter 14. Here we use the same setup as in Section 16.2.4, considering dynamic data for two-dimensional fields. Again, our data are the breakthrough times from a flow experiment, as the arrival times contain most of the available information.

In this section, we use only two resolutions, a coarse grid of 16×16 cells and a fine grid of 32×32 cells. Denote the observed breakthrough times by $\mathbf{p}_{obs}$, and note that this same dataset is used for all scales. Denote the unknown field of coarse log-permeabilities by $\mathbf{x}_0$ and the fine log-permeability field as $\mathbf{x}_1$. Let the function g represent the result of running a flow simulator (here S3D) on an input permeability field, so that $g(\mathbf{x}_0)$ is the vector of predicted breakthrough times for the coarse field $\mathbf{x}_0$ and $g(\mathbf{x}_1)$ is the vector of predicted breakthrough times for the fine field $\mathbf{x}_1$. We use independent Gaussian likelihoods at each scale:

$$
p(\mathbf{p}_{obs}|\mathbf{x}_0, \mathbf{x}_1, \tau_0, \tau_1) \propto \exp\left\{ - \frac{\tau_0}{2}(\mathbf{p}_{obs} - g(\mathbf{x}_0))'(\mathbf{p}_{obs} - g(\mathbf{x}_0)) \right.
$$
$$
\left. - \frac{\tau_1}{2}(\mathbf{p}_{obs} - g(\mathbf{x}_1))'(\mathbf{p}_{obs} - g(\mathbf{x}_1)) \right\}.
$$

The precisions τ_0 and τ_1 could be treated separately or could be forced to be equal.

We put a first-order symmetric intrinsic 2-D Markov random field prior on the coarse and fine scales. For each scale i,

$$p(\mathbf{x}_i|\beta_i) \propto \exp\left\{-\frac{\beta_i}{2}\mathbf{x}_i'\mathbf{H}_i\mathbf{x}_i\right\}, \tag{16.1}$$

where β_i is the smoothness parameter for that scale, and $\mathbf{H}$ is the precision matrix with diagonal elements equal to the number of neighbors of that grid cell and off-diagonal elements equal to -1 if the corresponding cells are adjacent and 0 otherwise. The priors for the different scales are treated as independent of each other. Finally, we specify informative gamma priors on τ_0, τ_1, β_0, and β_1.

Turning to the actual data, here we use the same Hill Air Force Base data as in Section 16.2.4. There are four injection wells along one short edge where water is pumped into the ground, and at a particular point in time a tracer is added so that its concentration can be monitored at other sites. Three production wells along the opposite edge extract water from the ground, forcing water to flow across the site. There are five sampling wells in the middle, where the concentration of the tracer is measured (see the upper left plot of Figure 16.10).

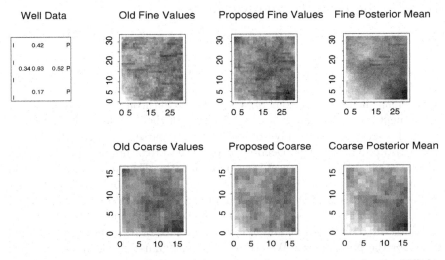

Fig. 16.10. Layout of wells, a sample swap, and posterior means for the Hill Air Force Base data. In the upper left plot, the wells are labeled "I" for injectors, "P" for producers, and the samplers are shown with numbers where the value is the breakthrough time (in days) for each well. For the permeability plots, all values are on the log scale, and lighter regions correspond to higher permeability values.

Running the Metropolis-coupling algorithm leads to good fits of the data. Figure 16.10 shows a representative swap between the coarse and fine scales,

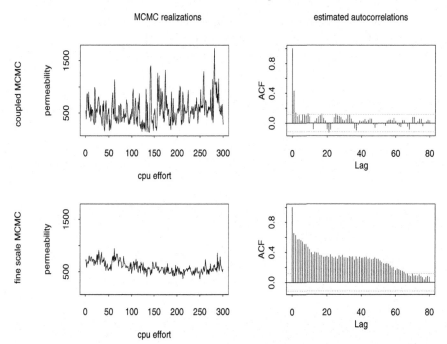

Fig. 16.11. MCMC trace plots and autocorrelation plots for the permeability at an interior pixel in the hydrology application under the coupled MCMC approach (top row) and standard MCMC within the fine scale only (bottom row). The trace plots are standardized to a comparable CPU time.

and one can see how the features of the fields are transferred between scales. Yet the resulting fields also have a reasonable amount of smoothness for their respective levels. The rightmost column of Figure 16.10 shows the posterior mean log-permeability fields, with the fine grid on top and the coarse grid beneath. Note that the lower and central sampling wells have the first and last breakthrough times, respectively. In order to be consistent with these data, the posterior permeability realizations need to contain a high-permeability path from the injectors to the lower sampling well, but also contain a rather abrupt decrease near the central sampling well. Our methods here have captured this structure. We note that such constraints can create severe difficulties in the mixing of MCMC in that fine-scale-only runs can have difficulties escaping from local modes. Coupling the two scales allows the coarse-scale sampler, which is far more efficient in exploring its posterior, to pass this improved performance on to the finer scale. Figure 16.11 shows MCMC trace plots and their associated autocorrelation functions for an interior pixel under the two sampling schemes. Notice the order-of-magnitude difference in the range of values explored by the coupled chain as compared with the single-scale chain.

Using Metropolis coupling yields autocorrelation times drastically smaller than those of the standard fine-scale MCMC algorithm.

Single Photon Emission Computed Tomography Example

Single photon emission computed tomography (SPECT) is a medical imaging technology. The patient is treated with a radioactive substance that releases individual gamma rays (in small amounts, so as to be detectable but not damaging to the patient). Some of these gamma rays are detected by a special camera, and then a classic inverse problem results from the need to infer the original image in the patient from the counts of detected gamma rays. Computer code links hypothesized images to predicted gamma ray counts. The primary goal is to reconstruct the features of the underlying object, such as a patient's brain. This image is typically desired at a particular resolution, and only one type of data is collected, so there is nothing inherently multiscale here. However, a fully Bayesian analysis requires the use of Markov chain Monte Carlo methods, and the chain can take a long time to run because of poor mixing. The use of multiscale techniques can greatly speed up the computational aspects of the problem, improving the statistical inference that can be done in a reasonable amount of time. Thus, we demonstrate the implicit methods of Part IV with this example.

Poor mixing in this example can result from the fact that the unknown of interest is an image, and naturally the pixels of the image are highly correlated. Adjacent pixels are more likely to have the same or close to the same values than are pixels farther away. In two or more dimensions, the correlations in images can make it difficult to efficiently explore the posterior. Furthermore, when computer simulators are involved, such as the SPECT simulator (or the flow simulator from Chapter 16), each likelihood evaluation is considerably more computationally expensive than in a standard statistical model. Thus, improving mixing has a serious impact on the overall computational efficiency.

The gamma cameras can be set up for either two-dimensional or three-dimensional imaging. Here we work only with the two-dimensional setup, but the approaches easily generalize to the three-dimensional case. Some imaging setups are to create a single static image, while others are designed to capture a series of images over time, allowing the creation of an imaging movie. We keep things simple by only dealing with a single static image here, as the need to handle both space and time requires a more sophisticated model to deal with different correlations in space and time. Of course, the multiscale methods directly generalize to the spatio-temporal setting when an appropriate model is used.

In this chapter, our primary goal is to demonstrate the multiscale methodology, and thus we work with a somewhat simplified version of the problem. We retain the key features of the real application but use several simplifying assumptions that could be dealt with in more sophisticated ways. For more complete treatments, we refer the reader to other references, such as Vardi et al. (1985) or Higdon et al. (1997).

Thus this chapter focuses on the problem of reconstructing a two-dimensional image. The goal is to build a photon emission intensity map of the underlying object by using photon counts detected by a gamma camera. Figure 17.1 shows a conceptualized version of the physical setup of the problem and the information available. The object is represented as a collection of pixels at a certain resolution. Each pixel emits photons at a certain rate x_i, where i indexes the pixel location, and the intensities are expected to vary by spatial location (which is what allows us to create the image). The gamma camera is really an array of detectors that count the number of photons observed. Lead columnators on the camera separate the detectors and absorb any photon that fails to hit the camera at nearly a right angle. The camera thus obtains counts within bins, with the bins indexed by b. The camera is fully rotated around the object so that photons can be observed from all directions. The angle of observation is denoted by a. The columnators ensure that the observed photons are from the correct angle. The observed data are the counts y_{ab} obtained from bin b of the gamma camera while it was positioned at angle a. Here we use 120 camera positions and 128 camera bins; i.e., $a \in \{1, \ldots, 120\}$ and $b \in \{1, \ldots, 128\}$.

A photon may be scattered, absorbed, miss the gamma camera, or otherwise fail to be detected. The columnators, along with the physical characteristics of the object, determine the probability, p_{abi}, that a photon emitted from pixel i is detected by camera bin b when the camera is positioned at angle a. Here we take the probabilities p_{abi} as known. In practice, prior calibration experiments are typically carried out to determine these probabilities, and they can then be taken as fixed and known.

It is then reasonable to model the counts y_{ab} as having a Poisson distribution with mean

$$\theta_{ab} = \sum_i x_i p_{abi}$$

or rewriting this in matrix form

$$\boldsymbol{\theta}_{n \times 1} = \mathbf{P}_{n \times m} \mathbf{x}_{m \times 1}, \tag{17.1}$$

where $n = 128 \times 120$ is the total number of observations and $m = 128 \times 128$ is the number of pixels. The object pixel intensities, x_i, are the unknowns of

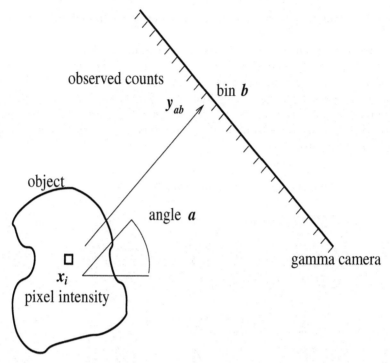

Fig. 17.1. An object emits photons with location-dependent intensities x_i. The gamma camera array produces counts of photon emissions detected within bins. The camera array is rotated around the object and at any given time of detection the observation angle is denoted by a. The counts from each angle a and each bin b are denoted y_{ab}.

interest, as the intensity map is the image that we wish to reconstruct. Given these intensities, the likelihood can be written as

$$L(\mathbf{y}|\mathbf{x}) \propto \prod_{a,b} \theta_{ab}^{y_{ab}} e^{-\theta_{ab}}. \tag{17.2}$$

More details on the derivation of this likelihood can be found in Vardi et al. (1985).

Primarily we are interested in this likelihood at this scale. However, it will be computationally helpful to consider an auxiliary coarser scale for the image intensity map of $m_0 = 64 \times 64$, which is half the resolution in each dimension. Note that the observed data $\mathbf{y}$ are the same regardless of the resolution for the underlying intensity map. Thus the likelihood is of the form (17.2) for all scales, where only the values of θ_{ab} may change with the change in scale. The resulting coarse mean vector $\boldsymbol{\theta}_0$ is still of length n, but it is the product of the $n \times m_0$-dimensional probability matrix $\mathbf{P}_0$ and the vectorized set of m_0 unknown coarse pixel intensities $\mathbf{x}_0$. So calculating the change in

θ_0 after changing an x_{0j} is four times faster as compared with the fine-scale calculation.

Because the likelihood is Poisson and the fine scale has exactly twice the resolution in each dimension, it makes sense to move between scales by aggregation. Thus the photon emission intensity of a coarse pixel is equal to the sum of the emission intensities of the four corresponding fine pixels. Note that this summation across scales is in contrast to the more common averaging to move from finer to coarser scales, and is chosen because of the particular physical details of the problem.

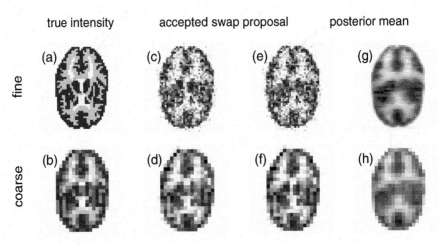

Fig. 17.2. Coupled fine- and coarse-scale MCMC for a SPECT example. (a) True emission intensities; (b) coarsened version of the true intensities; (c) and (d) current values for intensities $\mathbf{x}_1$ and $\mathbf{x}_0$ during the coupled MCMC run; (e) and (f) proposed fine and coarse images $\mathbf{x}^*$ and $\mathbf{x}_0^*$ after swapping an interior patch of the images in (c) and (d); (g) posterior mean for $\mathbf{x}_1$; (h) posterior mean for $\mathbf{x}_0$.

Here our data are from a computer-generated dataset from Higdon et al. (2002). The true image source intensity is shown in Figure 17.2(a). The corresponding coarsened version is given in Figure 17.2(b). Because there is a physical difference between pixels representing brain matter and pixels that are outside of the brain, we use a mask for moving between scales, so that only pixels within the brain area are dealt with in fully multiscale fashion. The pixels outside of the brain area are of less scientific interest, and emissions should be close to zero. Thus inference focuses only on the interior region. Figure 17.3 shows a schematic version, where the active region is shaded.

Following the analysis in Higdon et al. (2002), we use a symmetric first-order intrinsic Gaussian Markov random field (MRF) prior, as in Equation (14.5), for the intensities at each scale, $\mathbf{x}_1$ and $\mathbf{x}_0$. The precision matrices $\mathbf{H}_1$ and $\mathbf{H}_0$ have diagonal elements H_{ii} with values equal to the number of

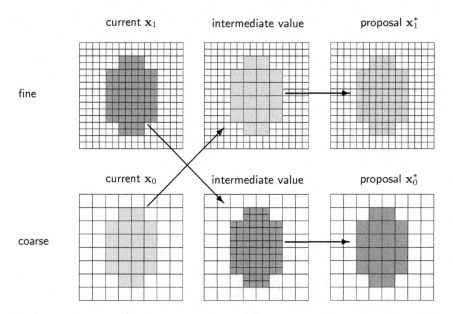

Fig. 17.3. A proposal that swaps only a piece of the image between the coarse and fine scales. Given the current values for x and x_0, the shaded regions of the two images are exchanged, giving the intermediate values. The coarse shaded piece is refined to give a fine proposal x^*, and the fine shaded piece is coarsened to give a coarse proposal x_0^*. The stochastic refining of the coarse shaded piece conditions on its previous coarse value as well as its neighboring fine-scale pixels.

neighboring sites; their off-diagonal elements H_{ij} are equal to -1 if sites i and j are adjacent and 0 otherwise. The MRF structure works well for multiresolution image analysis since it naturally occurs on a grid and because the methods of Chapters 14 and 15 are suitable for MRF models. Diffuse gamma priors are assigned to the β precision parameter for each scale. At a particular resolution, we perform standard MCMC, such as in Weir (1997) or Higdon et al. (1997).

17.1 Metropolis Coupling

Our first multiscale analysis uses the Metropolis coupling of Chapter 14, and in particular Section 14.2.2. For creating the Metropolis-Hastings proposal for moving from fine to coarse, the intensities in proposed coarse pixels are just the sum of the intensities of the four corresponding fine pixels. To get the proposed fine values from the current coarse image map, for each coarse pixel, the four corresponding fine pixels of the proposal are first set at a value which is one-quarter that of the coarse pixel, which produces the intermediate step in Figure 17.3. This is done for all of the proposed fine-scale pixels. After this step is complete, then the intermediate map is smoothed by taking each block of four fine pixels and drawing them from their conditional distribution, conditioning on them having the same mean value as their current value (or, equivalently, the same total value) and conditioned on the current values of all neighbors. This smoothing step helps create a more realistic proposal that has a reasonable probability of being accepted. The conditioning on the proper mean ensures reversibility. It is necessary to take this smoothing step into account when computing the Metropolis-Hastings ratio for the probability of acceptance.

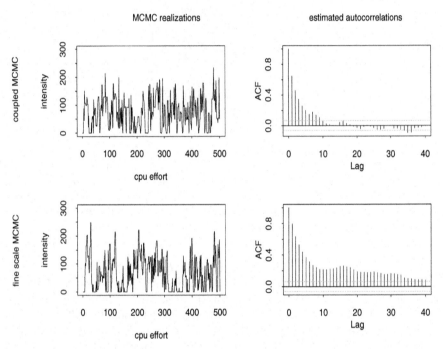

Fig. 17.4. Trace plots and autocorrelation plots for the intensity of an interior pixel under the Metropolis-coupled approach (top row) and standard single-scale MCMC (bottom row). The coupled MCMC is about three times as efficient when standardized to CPU effort.

For our SPECT example, this proposal has about a one in eight chance of acceptance. Figure 17.2 shows the before and after images for an accepted swap (images (c) to (f)) along with coarse- and fine-scale posterior mean images (images g and h). MCMC trace plots of the fitted intensity value of an interior pixel are shown in Figure 17.4 for Metropolis-coupled MCMC (top) and a single-scale MCMC run at the finer resolution only (bottom). The coupled MCMC produces notably better mixing as evidenced both in the trace plots and in the autocorrelation plots, where the estimated autocorrelation times (Sokal, 1987) for the coupled MCMC are about a third of those for the single-scale MCMC algorithm. Note that the trace plots are standardized to a comparable CPU time to account for the fact that Metropolis coupling is more computationally intensive. Yet this small additional effort clearly leads to overall gains in efficiency.

17.2 Genetic Algorithms

In this example, we can further improve upon the results from Metropolis coupling by using the multiscale MCMC methods derived from genetic algorithms in Chapter 15. We now add in the multiscale crossover swap proposals from Section 15.3.2 using each about half of the time.

For the crossover steps, we need to reparameterize so that the parameter vector is suitable for crossover sampling. In the notation of Chapter 15, let $\phi^{(2)} = \mathbf{x}^{(2)}$ and let $\phi^{(1)}$ be the sum of each block of four fine pixels corresponding to a single coarse pixel, so that $\phi^{(1)}$ and $\phi^{(2)}$ have the same dimension and interpretation. $\boldsymbol{\lambda}^{(1)}$ is then defined as the values of the upper left, upper right, and lower left pixels for each of the blocks of four fine pixels (since knowing those three and their sum determines the fourth pixel for each block). $\boldsymbol{\lambda}^{(1)}$ also contains $\beta^{(1)}$, and $\boldsymbol{\lambda}^{(2)} = \beta^{(2)}$. Finally, $\pi^{(i)}$ is the product of the diffuse gamma prior on $\beta^{(i)}$ and the MRF prior induced on $(\phi^{(i)}, \boldsymbol{\lambda}^{(i)})$ by Equation (14.5).

We then use both full swaps (the Metropolis-coupled swaps from above) and crossover swaps. While the full swaps exchange all of the ϕ values (i.e., the full grid), the crossover swaps exchange only a subset of ϕ values and sample only the associated pixels in $\boldsymbol{\lambda}^{(1)}$, as well as $\beta^{(1)}$ and $\beta^{(2)}$. (An alternative crossover scheme could swap clusters of coarse pixels, such as a 4×4 block, which might preserve more of the local structure of the MRF prior.)

About one in eight full swaps and about one in five crossover swaps were accepted. Crossover swaps often have larger acceptance probabilities because less of the parameter space is being changed at the same time, yet they still provide a large increase in mixing when compared with most single-scale techniques. However, they do not improve mixing as much as an accepted full swap typically does since it changes the entire parameter space. Thus the combination of full and crossover swaps can be quite beneficial. Figure 17.5 shows an accepted full swap along with coarse- and fine-scale posterior mean images. Autocorrelation functions are shown in Figure 17.6 for a representative fine pixel under three posterior sampling schemes: a single-resolution chain, a multiresolution approach that only uses Metropolis-coupled full swaps, and the combination of both full and crossover swaps. Comparing the two multiscale results provides a method for attempting to separate out the effect of using multiple scales from the additional gains made by the genetic algorithm-style

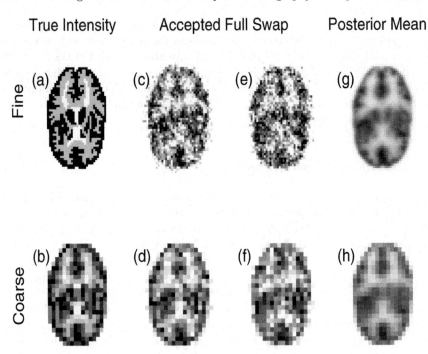

Fig. 17.5. Modeling using multiscale genetic algorithm-style MCMC. (a) True emission intensities; (b) coarsened version of the true intensities; (c) and (d) snapshot of current fine and coarse intensity values right before a swap; (e) and (f) fine and coarse images after a swap; (g) fine posterior mean intensity map; (h) coarse posterior mean intensity map.

proposals. The combination approach reduces the estimated autocorrelation times by more than Metropolis coupling alone does, even after adjusting for comparable CPU time. As such, gains from any multiscale approach are significant, and additional gains can be obtained from efficient multiscale algorithms such as this combination swapping approach.

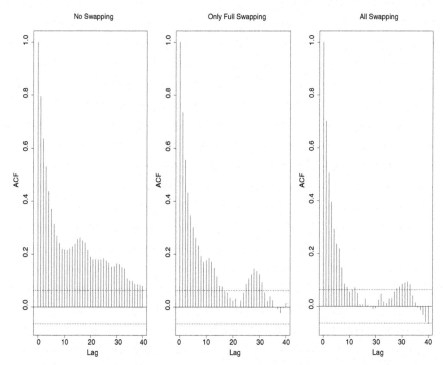

Fig. 17.6. Autocorrelation plots for the intensity of an interior pixel in the SPECT application under the single-chain standard MCMC at the fine scale only (left), multiscale with full swaps only (middle), and our multiscale MCMC approach with full and crossover swaps (right).

18

Conclusions

Multiscale modeling presents a number of challenges beyond standard statistical modeling. In particular, there is the need for consistent modeling across different scales. The methods presented in this book represent a wide variety of such approaches, having been grouped into three general categories: methods for building a model at the finest levels that can be simplified for coarser scales, methods for explicitly modeling the relationship between scales, and methods that implicitly link the scales. We expect that additional new approaches will continue to be developed in the future.

Depending on the particular problem, some approaches may be more relevant or sensible than others. In some cases, the analyst will have a choice of possible multiscale approaches. As many of these methods have only been developed quite recently, there has been little analysis comparing the methods, either analytically or empirically. We want to encourage this investigation, and hope that such research will produce practical suggestions and guidelines.

The Bayesian approach is powerful, allowing a number of approaches that would be difficult or impossible without the ability to incorporate prior information. It also allows a coherent accounting of uncertainty, something that can be quite difficult to do across multiple scales. Markov chain Monte Carlo methods provide a general mechanism for fitting these models.

In many cases, the multiscale problem is difficult because of a lack of direct information for moving between scales. Care must be taken not to incorporate incorrect information, which can easily sneak in when specifying priors. It is common to use priors that are convenient either because they are conjugate (or conditionally conjugate) or because a particular parametric form makes the specification of hyperparameters easier. Such practices can have inadvertent effects on the posterior. As multiscale problems are often characterized by not having as much data as one would like at all of the scales, the prior information becomes quite important. This setting necessitates serious thinking about the choice of priors. It would also be a good idea to conduct a sensitivity analysis to see how much the posterior depends on the choice of the prior. In many cases, it will be highly dependent, but one can see if a reasonable range of prior specifications leads to similar conclusions, and one can see which parameters have the most influence and hence which parts of the prior are the most important.

References

Abrahamsen, P. (1997). "A review of Gaussian random fields and correlation functions." Technical Report 917, Norwegian Computing Center.

Abramovich, F., Sapatinas, T., and Silverman, B. W. (1998). "Wavelet thresholding via a Bayesian approach." *Journal of the Royal Statistical Society, Series B*, 60, 725–749.

Abramowitz, M. and Stegun, I. A. (1964). *Handbook of Mathematical Functions with Formulas, Graphs, and Mathematical Tables.* 9th ed. New York: Dover.

Albert, J. H. and Chib, S. (1993). "Bayes inference via Gibbs sampling of autoregressive time series subjec to to Markov mean and variance shifts." *Journal of Business and Economic Statistics*, 11, 1–15.

Amemiya, T. and Wu, R. Y. (1972). "The effect of aggregation on prediction in the autoregressive model." *Journal of the American Statistical Association*, 67, 628–632.

Annable, M. D., Rao, P. S. C., Hatfield, K., Graham, W. D., Wood, A. L., and Enfield, C. G. (1998). "Partitioning tracers for measuring residual NAPL: Field-scale test results." *Journal of Environmental Engineering*, 124, 498–503.

Bäck, T. (1993). "Optimal mutation rates in genetic search." In *Proceedings of the Fifth International Conference on Genetic Algorithms*, ed. S. Forrest, 2–8. San Mateo, CA: Morgan Kaufmann.

Banerjee, S., Carlin, B. P., and Gelfand, A. E. (2003). *Hierarchical Modeling and Analysis for Spatial Data.* Boca Raton, FL: Chapman and Hall.

Barone, P., Sebastiani, G., and Stander, J. (2002). "Over-relaxation methods and coupled Markov chains for Monte Carlo simulation." *Statistics and Computing*, 12, 17–26.

Barry, R. P. and Ver Hoef, J. M. (1996). "Blackbox Kriging: Spatial prediction without specifying variogram models." *Journal of Agricultural, Biological, and Environmental Statistics*, 1, 297–322.

Basseville, M., Benveniste, A., and Willsky, A. S. (1992a). "Multiscale autoregressive processes, part I: Schur-Levinson parametrizations." *IEEE Transactions on Signal Processing*, 40, 1915–1934.

— (1992b). "Multiscale autoregressive processes, part II: Lattice structures for whitening and modeling." *IEEE Transactions on Signal Processing*, 40, 1935–1953.

Beran, J. (1994). *Statistical Methods for Long Memory Processes*. Boca Raton, FL: Chapman and Hall.

Berger, J. O., de Oliveira, V., and Sansó, B. (2001). "Objective Bayesian analysis of spatially correlated data." *Journal of the American Statistical Association*, 96, 1361–1374.

Besag, J. (1974). "Spatial interaction and the statistical analysis of lattice systems (with discussion)." *Journal of the Royal Statistical Society, Series B*, 36, 192–236.

— (2000). "Markov chain Monte Carlo for statistical inference." Technical Report, University of Washington, Center for Statistics and the Social Sciences.

Besag, J., Green, P., Higdon, D., and Mengersen, K. (1995). "Bayesian computation and stochastic systems (with discussion)." *Statistical Science*, 10, 3–66.

Besag, J. and Kooperberg, C. (1995). "On conditional and intrinsic autoregressions." *Biometrika*, 82, 733–746.

Besag, J., York, J., and Mollié, A. (1991). "Bayesian image restoration, with two applications in spatial statistics (with discussion)." *Annals of the Institute of Statistical Mathematics*, 43, 1–59.

Best, N. G., Ickstadt, K., and Wolpert, R. L. (2000). "Spatial Poisson regression for health and exposure data measured at disparate resolutions." *Journal of the American Statistical Association*, 95, 1076–1088.

Bollerslev, T. and Wright, J. H. (2000). "Semiparametric estimation of long-memory volatility dependencies: The role of high-frequency data." *Journal of Econometrics*, 98, 81–106.

Bonet-Cunha, L., Oliver, D. S., Redner, R. A., and Reynolds, A. C. (1998). "A hybrid Markov chain Monte Carlo method for generating permeability fields conditioned to multiwell pressure data and prior information." *SPE Journal*, 3, 261–271.

Bouman, C. and Liu, B. (1991). "Multiple resolution segmentation of textured images." *IEEE Transactions on Pattern Analysis and Machine Intelligence*, 13, 99–113.

Bouman, C. A. and Shapiro, M. (1994). "A multiscale random field model for Bayesian image segmentation." *IEEE Transactions on Image Processing*, 3, 162–177.

Bouman, P., Dukic, V., and Mengu, X.-L. (2005). "A Bayesian multiresolution hazard model with application to an AIDS reporting delay study." *Statistica Sinica*, 15, 325–357.

Breiman, L., Friedman, J. H., Olshen, R., and Stone, C. (1984). *Classification and Regression Trees*. Belmont, CA: Wadsworth.

Briggs, W. L., Henson, V. E., and McCormick, S. F. (2000). *A Multigrid Tutorial*. 2nd ed. Philadelphia: Society for Industrial and Applied Mathematics.

Brockwell, P. J. and Davis, R. A. (1991). *Time Series: Theory and Methods*. 2nd ed. New York: Springer.

Brooks, S. P. (1998). "Markov chain Monte Carlo method and its application." *The Statistician*, 47, 69–100.

Brown, P. J., Fearn, T., and Vannucci, M. (2001). "Bayesian wavelet regression on curves with application to a spectroscopic calibration problem." *Journal of the American Statistical Association*, 96, 398–408.

Brownie, C., Bowman, D. T., and Burton, J. W. (1994). "Estimating spatial variation in analysis of data from yield trials: A comparison of methods." *Agronomy Journal*, 85, 1244–1253.

Bustamante, C., Boaventura, E. G., Tassinari, W., Reis, M. G., Carvalho, M., and Ko, A. I. (2006). "The effect of climate on cyclic epidemic transmission of urban leptospirosis." Technical Report, Fiocruz.

Calder, C. A., Holloman, C., and Higdon, D. M. (2002). "Exploring space-time structure in ozone concentration using a dynamic process convolution model." In *Case Studies in Bayesian Statistics 6*, eds. C. Gatsonis, R. E. Kass, A. Carriquiry, A. Gelman, D. Higdon, D. K. Pauler, and I. Verdinelli, 165–176. New York: Springer.

Calderón, A. P. (1964). "Intermediate spaces and interpolation, the complex method." *Studia Mathematica*, 24, 113–190.

Carlin, B. P. and Banerjee, S. (2003). "Hierarchical multivariate CAR models for spatio-temporally correlated survival data." In *Bayesian Statistics 7*, eds. J. M. Bernardo, M. J. Bayarri, J. O. Berger, A. P. Dawid, D. Heckerman, A. F. M. Smith, and M. West, 45–63. Oxford: Oxford University Press.

Carlin, B. P. and Chib, S. (1995). "Bayesian model choice via Markov chain Monte Carlo methods." *Journal of the Royal Statistical Society, Series B*, 57, 473–484.

Carter, C. K. and Kohn, R. (1994). "On Gibbs sampling for state space models." *Biometrika*, 81, 541–553.

Chatterjee, S., Laudato, M., and Lynch, L. A. (1996). "Genetic algorithms and their statistical applications: An introduction." *Computational Statistics and Data Analysis*, 22, 633–651.

Chib, S. and Greenberg, E. (1994). "Bayes inference in regression models with ARMA(p,q) errors." *Journal of Econometrics*, 64, 183–206.

Chilès, J.-P. and Delfiner, P. (1999). *Geostatistics: Modeling Spatial Uncertainty*. New York: John Wiley and Sons.

Chipman, H., George, E., and McCulloch, R. (1998). "Bayesian CART model search (with discussion)." *Journal of the American Statistical Association*, 93, 935–960.

Chipman, H. A., George, E. I., and McCulloch, R. E. (2002). "Bayesian treed models." *Machine Learning*, 48, 303–324.

Chipman, H. A., Kolaczyk, E. D., and McCulloch, R. E. (1997). "Adaptive Bayesian wavelet shrinkage." *Journal of the American Statistical Association*, 92, 1413–1421.

Chopin, N. and Pelgrin, F. (2004). "Bayesian inference and state number determination for hidden Markov models: An application to the information content of the yield curve about inflation." *Journal of Econometrics*, 123, 327–344.

Chou, K. C., Willsky, A. S., and Benveniste, A. (1994a). "Multiscale recursive estimation, data fusion, and regularization." *IEEE Transactions on Automatic Control*, 39, 464–478.

Chou, K. C., Willsky, A. S., and Nikoukhah, R. (1994b). "Multiscale systems, Kalman filters, and Riccati equations." *IEEE Transactions on Automatic Control*, 39, 479–492.

Clyde, M. and George, E. I. (2000). "Flexible empirical Bayes estimation for wavelets." *Journal of the Royal Statistical Society, Series B*, 62, 681–698.

Clyde, M., Parmigiani, G., and Vidakovic, B. (1998). "Multiple shrinkage and subset selection in wavelets." *Biometrika*, 85, 391–402.

Cohen, A., Daubechies, I., Jawerth, B., and Vial, P. (1993). "Multiresolution analysis, wavelets and fast algorithms on an interval." *Comptes Rendus I*, 316, 417–421.

Comer, M. and Delp, E. (1999). "Segmentation of textured images using a multiresolution Gaussian autoregressive model." *IEEE Transactions on Image Processing*, 8, 408–420.

Craig, P. S., Goldstein, M., Seheult, A. H., and Smith, J. A. (1996). "Bayes linear strategies for history matching of hydrocarbon reservoirs." In *Bayesian Statistics 5*, eds. J. M. Bernardo, J. O. Berger, A. P. Dawid, and A. F. M. Smith, 69–95. Oxford: Clarendon Press.

Cressie, N. A. C. (1993). *Statistics for Spatial Data, revised edition*. New York: John Wiley and Sons.

Crouse, M. S., Nowak, R. D., and Baraniuk, R. G. (1998). "Wavelet-based statistical signal processing using hidden Markov models." *IEEE Transactions on Signal Processing*, 46, 886–902.

Damian, D., Sampson, P. D., and Guttorp, P. (2001). "Bayesian estimation of semiparametric nonstationary spatial covariance structure." *Environmetrics*, 12, 161–178.

Daoudi, K., Frakt, A. B., and Willsky, A. S. (1999). "Multiscale autoregressive models and wavelets." *IEEE Transactions on Information Theory*, 45, 828–845.

Daubechies, I. (1992). *Ten Lectures on Wavelets*. Philadelphia: Society for Industrial and Applied Mathematics.

Dawid, A. P. (1992). "Applications of a general propagation algorithm for probabilistic expert systems." *Statistics and Computing*, 2, 25–36.

DeJong, K. (1975). "An Analysis of the Behavior of a Class of Genetic Adaptive Systems." Ph.D. thesis, University of Michigan, Ann Arbor, MI.

Dempster, A. P., Laird, N., and Rubin, D. B. (1977). "Maximum likelihood from incomplete data via the EM algorithm." *Journal of the Royal Statistical Society, Series B*, 39, 1–38.

Dempster, A. P., Selwyn, M. R., Patel, C. M., and Roth, A. J. (1984). "Statistical and computational aspects of mixed model analysis." *Applied Statistics*, 33, 203–214.

Denison, D. G. T., Holmes, C. C., Mallick, B. K., and Smith, A. F. M. (2002). *Bayesian Methods for Nonlinear Classification and Regression*. London: John Wiley and Sons.

Denison, D. G. T., Mallick, B. K., and Smith, A. F. M. (1998). "A Bayesian CART algorithm." *Biometrika*, 85, 363–377.

Diaconis, P. and Zabell, S. L. (1982). "Updating subjective probability." *Journal of the American Statistical Association*, 77, 822–830.

Donoho, D. L. and Johnstone, I. M. (1994). "Ideal spatial adaptation via wavelet shrinkage." *Biometrika*, 81, 425–455.

— (1995). "Adapting to unknown smoothness via wavelet shrinkage." *Journal of the American Statistical Association*, 90, 1200–1224.

Donoho, D. L., Johnstone, I. M., Kerryacharian, G., and Picard, D. (1995). "Wavelet shrinkage: Asymptopia?" *Journal of the Royal Statistical Society, Series B*, 57, 301–369.

Drost, F. C. and Nijman, T. E. (1993). "Temporal aggregation of GARCH processes." *Econometrica*, 61, 909–927.

Evans, M. and Swartz, T. (1995). "Methods for approximating integrals in statistics with special emphasis on Bayesian integration problems (Disc: V11 P54-64)." *Statistical Science*, 10, 254–272.

Fang, K.-T., Li, R., and Sudjianto, A. (2006). *Design and Modeling for Computer Experiments*. Boca Raton, FL: Chapman and Hall/CRC.

Ferreira, M. A. R., Bi, Z., West, M., Lee, H., and Higdon, D. (2003). "Multiscale modelling of 1-D permeability fields." In *Bayesian Statistics 7*, eds. J. M. Bernardo, M. J. Bayarri, J. O. Berger, A. P. Dawid, D. Heckerman, A. F. M. Smith, and M. West, 519–527. Oxford: Oxford University Press.

Ferreira, M. A. R. and de Oliveira, V. (2007). "Bayesian reference analysis for Gaussian Markov random fields." *Journal of Multivariate Analysis*, 98, 789–812.

Ferreira, M. A. R., Higdon, D., Lee, H. K. H., and West, M. (2005). "Multiscale random field models." Technical Report, UFRJ - DME.

Ferreira, M. A. R., West, M., Lee, H. K. H., and Higdon, D. (2006). "Multiscale and hidden resolution time series models." *Bayesian Analysis*, 1, 947–968.

Floris, F. J. T., Bush, M. D., Cuypers, M., Roggero, F., and Syversveen, A.-R. (2001). "Methods for quantifying the uncertainty of production forecasts: A comparative study." *Petroleum Geosciences*, 7, S87–S96.

Flowerdew, R. and Green, M. (1989). "Statistical methods for inference between incompatible zonal systems." In *Accuracy of Spatial Databases*, eds. M. Goodchild and S. Gopal, 239–247. London: Taylor and Francis.

— (1994). "Areal interpolation and types of data." In *Spatial Analysis and GIS*, eds. S. Fotheringham and P. Rogerson, 121–145. London: Taylor and Francis.

Forney, G. D. (1973). "The Viterbi algorithm." *Proceedings of the IEEE*, 61, 268–278.

Frakt, A. B. and Willsky, A. S. (1998). "Multiscale autoregressive models and the stochastic realization problem." In *32nd Asilomar Conference on Signals, Systems and Computers*, 747–751. Pacific Grove, CA: IEEE.

Frühwirth-Schnatter, S. (1994). "Data augmentation and dynamic linear models." *Journal of Time Series Analysis*, 15, 183–202.

Fuentes, M. and Smith, R. L. (2001). "A new class of nonstationary spatial models." Technical Report 2534, North Carolina State University, Department of Statistics.

Gamerman, D. and Lopes, H. F. (2006). *Markov Chain Monte Carlo: Stochastic Simulation for Bayesian Inference*. 2nd ed. Boca Raton, FL: Chapman and Hall/CRC.

Gehlke, C. E. and Biehl, K. (1934). "Certain effects of grouping upon the size of the correlation coefficient in census tract material." *Journal of the American Statistical Association*, 29, 169–170.

Gelfand, A. E., Schmidt, A. M., Banerjee, S., and Sirmans, C. F. (2004). "Nonstationary multivariate process modeling through spatially varying coregionalization (with discussion)." *Test*, 13, 1–50.

Gelhar, L. W. (1993). *Stochastic Subsurface Hydrology*. Englewood Cliffs, NJ: Prentice-Hall.

Gelman, A. (1996). "Inference and monitoring convergence." In *Markov Chain Monte Carlo in Practice*, eds. W. R. Gilks, S. Richardson, and D. J. Spiegelhalter, 131–143. London: Chapman and Hall.

Gelman, A., Carlin, J. B., Stern, H. S., and Rubin, D. B. (1995). *Bayesian Data Analysis*. London: Chapman and Hall.

Geyer, C. J. (1991). "Markov chain Monte Carlo maximum likelihood." In *Computing Science and Statistics. Proceedings of the 23rd Symposium on the Interface*, 156–163. Fairfax Station, VA: Interface Foundation of North America.

Geyer, C. J. and Thompson, E. A. (1995). "Annealing Markov chain Monte Carlo with applications to ancestral inference." *Journal of the American Statistical Association*, 90, 909–920.

Gidas, B. (1989). "A renormalization group approach to image processing problems." *IEEE Transactions on Pattern Analysis and Machine Intelligence*, 11, 164–180.

Goldberg, D. E., Korb, B., and Deb, K. (1989). "Messy genetic algorithms: Motivation, analysis, and first results." *Complex Systems*, 3, 493–530.

Goodman, J. and Sokal, A. D. (1989). "Multigrid Monte Carlo method." *Physical Review Letters D*, 40, 2035–2072.

Gotway, C. A. and Young, L. J. (2002). "Combining incompatible spatial data." *Journal of the American Statistical Association*, 97, 632–648.

Gramacy, R. B. and Lee, H. K. H. (2006). "Bayesian treed Gaussian process models." Technical Report ams2006-08, Dept. of Applied Math and Statistics, University of California, Santa Cruz.

Green, P. J. (1995). "Reversible jump Markov chain Monte Carlo computation and Bayesian model determination." *Biometrika*, 82, 711–732.

Grossmann, A. and Morlet, J. (1984). "Decomposition of Hardy Functions into Square Integrable Wavelets of Constant Shape." *SIAM Journal on Mathematical Analysis*, 15, 723–736.

Haar, A. (1910). "Zur theorie der orthogonalen funktionensysteme." *Mathematische Annalen*, 69, 331–371.

Haas, T. C. (1990). "Lognormal and moving window methods of estimating acid deposition." *Journal of the American Statistical Association*, 85, 950–963.

Hamilton, J. D. (1989). "A new approach to the economic analysis of nonstationary time series and the business cycle." *Econometrica*, 57, 357–384.

Harvey, A. C. (1989). *Forecasting, Structural Time Series Models and the Kalman Filter*. Cambridge: Cambridge University Press.

Harville, D. A. (1997). *Matrix Algebra from a Statistician's Perspective*. New York: Springer-Verlag.

Hastie, T., Tibshirani, R., and Friedman, J. (2001). *The Elements of Statistical Learning*. New York: Springer-Verlag.

Hastings, W. K. (1970). "Monte Carlo sampling methods using Markov chains and their applications." *Biometrika*, 57, 97–109.

Hegstad, B. K. and Omre, H. (1997). "Uncertainty assessment in history matching and forecasting." In *Geostatistics Wollongong '96, Vol. 1.*, ed. E. Y. Baafi and N. A. Schofield, 585–596. Dordrecht: Kluwer Academic Publishers.

Higdon, D. (2002). "Space and space-time modeling using process convolutions." In *Quantitative Methods for Current Environmental Issues*, eds. C. Anderson, V. Barnett, P. C. Chatwin, and A. H. El-Shaarawi, 37–56. London: Springer-Verlag.

Higdon, D., Lee, H., and Bi, Z. (2002). "A Bayesian approach to characterizing uncertainty in inverse problems using coarse and fine scale information." *IEEE Transactions on Signal Processing*, 50, 389–399.

Higdon, D. M., Johnson, V. E., Bowsher, J. E., Turkington, T. G., Gilland, D. R., and Jaszczack, R. J. (1997). "Fully Bayesian estimation of Gibbs hyperparameters for emission computed tomography data." *IEEE Transactions on Medical Imaging*, 16, 516–526.

Higdon, D. M., Lee, H., and Holloman, C. (2003). "Markov chain Monte Carlo-based approaches for inference in computationally intensive inverse problems." In *Bayesian Statistics 7, Proceedings of the Seventh Valencia*

International Meeting, eds. J. M. Bernardo, M. J. Bayarri, J. O. Berger, A. P. Dawid, D. Heckerman, A. F. M. Smith, and M. West. Oxford: Oxford University Press.

Higdon, D. M., Swall, J., and Kern, J. C. (1999). "Non-stationary spatial modeling." In *Bayesian Statistics 6*, eds. J. M. Bernardo, J. O. Berger, A. P. Dawid, and A. F. M. Smith, 761–768. Oxford: Oxford University Press.

Hjort, N. L. and Omre, H. (1994). "Topics in spatial statistics." *Scandinavian Journal of Statistics*, 21, 289–357.

Holland, J. H. (1975). *Adaptation in Natural and Artificial Systems*. Ann Arbor: The University of Michigan Press.

Holloman, C. (2002). "Parameter Estimation Algorithms for Computationally Intensive Spatial Problems." Ph.D. thesis, Duke University, Durham, NC.

Holloman, C. H., Lee, H. K. H., and Higdon, D. M. (2006). "Multi-resolution genetic algorithms and Markov chain Monte Carlo." *Journal of Computational and Graphical Statistics*, 15, 861–879.

Holmes, C. C. and Mallick, B. K. (1998). "Parallel Markov chain Monte Carlo sampling." Technical Report, Imperial College, London.

Host, G., Omre, H., and Switzer, P. (1995). "Spatial interpolation errors for monitoring data." *Journal of the American Statistical Association*, 90, 853–861.

Houghton, J. T., Jenkins, G. J., and Ephraums, J. J., eds. (1990). *Climate Change: The IPCC Scientific Assessment. Intergovernmental Panel on Climate Change*. Cambridge: Cambridge University Press.

Huang, H.-C. and Cressie, N. (1997). "Multiscale spatial modeling." In *ASA Proceedings of the Section on Statistics and the Environment*, 49–54. Alexandria, VA: American Statistical Association.

— (2001). "Multiscale graphical modeling in space: applications to command and control." In *Spatial Statistics: Methodological Aspects and Applications*, ed. M. Moore, 83–113. New York: Springer.

Huang, H.-C., Cressie, N., and Gabrosek, J. (2002). "Fast, resolution-consistent spatial prediction of global processes from satellite data." *Journal of Computational and Graphical Statistics*, 11, 63–88.

Husmeier, D. and McGuire, G. (2003). "Detecting recombination in 4-taxa DNA sequence alignments with Bayesian hidden Markov models and Markov chain Monte Carlo." *Molecular Biology and Evolution*, 20, 315–337.

Hwang, S. (2000). "The effects of systematic sampling and temporal aggregation on discrete time long memory processes and their finite sample properties." *Econometric Theory*, 16, 347–372.

Ickstadt, K. and Wolpert, R. L. (1999). "Spatial regression for marked point processes." In *Bayesian Statistics 6*, eds. J. M. Bernardo, J. O. Berger, A. P. Dawid, and A. F. M. Smith, 323–341. Oxford: Oxford University Press.

Irving, W. W., Fieguth, P. W., and Willsky, A. S. (1997). "An overlapping tree approach to multiscale stochastic modeling and estimation." *IEEE Transactions on Image Processing*, 6, 1517–1529.

Isaaks, E. H. and Srivastava, R. M. (1989). *Applied Geostatistics*. Oxford: Oxford University Press.

James, A. I., Graham, W. D., Hatfield, K., Rao, P. S. C., and Annable, M. D. (1997). "Optimal estimation of residual non-aqueous phase liquid saturation using partitioning tracer concentration data." *Water Resources Research*, 33, 2621–2636.

Jeffrey, R. C. (1988). "Conditioning, kinematics, and exchangeability." In *Causation, Chance, and Credence*, eds. B. Skyrms and W. L. Harper, vol. 1, 221–255. Dordrecht: Kluwer.

— (1992). *Probability and the Art of Judgement*. New York: Cambridge University Press.

Jin, M., Delshad, M., Dwarakanath, V., McKinney, D. C., Pope, G. A., Sepehrnoori, K., Tilburg, C. E., and Jackson, R. E. (1995). "Partitioning tracer test for detection, estimation, and remediation performance assessment of subsurface non-aqueous phase liquids." *Water Resources Research*, 31, 1201–1211.

Johnson, V. E. (1998). "A coupling-regeneration scheme for diagnosing convergence in Markov chain Monte Carlo algorithms." *Journal of the American Statistical Association*, 93, 238–248.

Johnstone, I. M. and Silverman, B. W. (1998). "Empirical Bayes approaches to mixture problems and wavelet regression." Technical Report, Department of Mathematics, University of Bristol.

— (2004). "Needles and straw in haystacks: Empirical Bayes estimates of possibly sparse sequences." *Annals of Statistics*, 32, 1594–1649.

— (2005). "Empirical Bayes selection of wavelet thresholds." *Annals of Statistics*, 33, 1700–1752.

Journel, A. G. and Huijbregts, C. J. (1978). *Mining Geostatistics*. New York: Academic Press.

Juang, B. H. and Rabiner, L. R. (1991). "Hidden Markov models for speech recognition." *Technometrics*, 33, 251–272.

Kato, Z., Berthod, M., and Zerubia, J. (1996a). "A hierarchical Markov random field model and multi-temperature annealing for parallel image classification." *Graphical Models and Image Processing*, 58, 18–37.

Kato, Z., Zerubia, J., and Berthod, M. (1996b). "Unsupervised parallel image classification using Markovian models." *Graphical Models and Image Processing*, 58, 18–37.

Katul, G. G., Vidakovic, B., and Albertson, J. D. (2001). "Estimating global and local scaling exponents in turbulent flows using wavelet transformations." *Physics of Fluids*, 13, 241–250.

Kennedy, M. C. and O'Hagan, A. (2000). "Predicting the output from a complex computer code when fast approximations are available." *Biometrika*, 87, 1–13.

— (2001). "Bayesian calibration of computer models." *Journal of the Royal Statistical Society, Series B*, 63, 425–464.

Kern, J. C. (2000). "Bayesian Process-Convolution Approaches to Specifying Spatial Dependence Structure." Ph.D. thesis, Duke University, Durham, NC.

Kim, H.-M., Mallick, B. K., and Holmes, C. C. (2005). "Analyzing nonstationary spatial data using piecewise Gaussian processes." *Journal of the American Statistical Association*, 100, 653–668.

Kim, S. S., Reddy, A. L. N., and Vannucci, M. (2004). "Detecting traffic anomalies using discrete wavelet transform." In *Proceedings of the International Conference on Information Networking*, eds. H. K. Kahng and S. Goto, 951–961. Berlin: Springer-Verlag.

King, M. J. and Datta-Gupta, A. (1998). "Streamline simulation: A current perspective." *In Situ*, 22, 91–140.

Kirkpatrick, S., Gelatt, C. D., and Vecchi, M. P. (1983). "Optimization by simulated annealing." *Science*, 220, 671–680.

Ko, A. I., Reis, M. G., Dourado, C. R., Johnson, W. D., and Riley, L. W. (1999). "Urban epidemic of severe Leptospirosis in Brazil. Salvador Leptospirosis Study Group." *Lancet*, 354, 820–825.

Ko, K. and Vannucci, M. (2006a). "Bayesian wavelet analysis of autoregressive fractionally integrated moving-average processes." *Journal of Statistical Planning and Inference*, 136, 3415–3434.

— (2006b). "Bayesian wavelet-based methods for the detection of multiple changes of the long memory parameter." *IEEE Transactions on Signal Processing*, 54, 4461–4470.

Kolaczyk, E. D. (1999). "Bayesian multiscale models for Poisson processes." *Journal of the American Statistical Association*, 94, 920–933.

Kolaczyk, E. D. and Huang, H. (2001). "Multiscale statistical models for hierarchical spatial aggregation." *Geographical Analysis*, 33, 95–118.

Kolaczyk, E. D., Ju, J., and Gopal, S. (2005). "Multiscale, multigranular statistical image segmentation." *Journal of the American Statistical Association*, 100, 1358–1369.

Krishnamachari, S. and Chellappa, R. (1997). "Multiresolution Gauss-Markov random field models for texture segmentation." *IEEE Transactions on Image Processing*, 6, 251–267.

Kwon, D. W., Ko, K., Vannucci, M., Reddy, A. L. N., and Kim, S. (2006). "Wavelet methods for the detection of anomalies and their application to network traffic analysis." *Quality and Reliability Engineering International*, 22, 1–17.

Laferté, J.-M., Heitz, F., Pérez, P., and Fabre, E. (1995). "Hierarchical statistical models for the fusion of multiresolution image data." In *Proceedings of the International Conference on Computer Vision*, 908. Washington, DC: IEEE Computer Society.

Laferté, J.-M., Pérez, P., and Heitz, F. (2000). "Discrete Markov image modeling and inference on the quadtree." *IEEE Transactions on Image Processing*, 9, 390–404.

Lakshmanan, S. and Derin, H. (1993). "Gaussian Markov random fields at multiple resolutions." In *Markov Random Fields: Theory and Applications*, eds. R. Chellapa and A. Jain, 131–157. New York: Academic Press.

Lee, H. K. H. (2004). *Bayesian Nonparametrics via Neural Networks*. ASA-SIAM Series on Statistics and Applied Probability. Philadelphia: Society for Industrial and Applied Mathematics.

Lee, H. K. H., Higdon, D., Bi, Z., Ferreira, M. A. R., and West, M. (2002). "Markov random field models for high-dimensional parameters in simulations of fluid flow in porous media." *Technometrics*, 44, 230–241.

Lee, H. K. H., Higdon, D. M., Calder, C. A., and Holloman, C. H. (2005). "Efficient models for correlated data via convolutions of intrinsic processes." *Statistical Modelling*, 5, 53–74.

Liang, F. (2003). "Use of sequential structure in simulation from high-dimensional systems." *Physical Review E*, 67, 056101-1–7.

Liang, F. and Wong, W. H. (2001). "Real parameter evolutionary Monte Carlo with applications to Bayesian mixture models." *Journal of the American Statistical Association*, 96, 653–666.

Liò, P. and Vannucci, M. (2000). "Wavelet change-point prediction of transmembrane proteins." *Bioinformatics*, 16, 376–382.

Liu, J. S., Neuwald, A. F., and Lawrence, C. E. (1999). "Markovian structures in biological sequence alignments." *Journal of the American Statistical Association*, 94, 1–15.

Liu, J. S. and Sabatti, C. (1999). "Simulated sintering: Markov chain Monte Carlo with spaces of varying dimensions (with discussion)." In *Bayesian Statistics 6*, eds. J. M. Bernardo, J. O. Berger, A. P. Dawid, and A. F. M. Smith, 389–413. Oxford: Oxford University Press.

— (2000). "Generalised Gibbs sampler and multigrid Monte Carlo for Bayesian computation." *Biometrika*, 87, 353–369.

Loschi, R. H., Iglesias, P. L., and Arellano-Valle, R. B. (2002). "Conditioning on uncertain event: Extensions to Bayesian inference." *Test*, 11, 1–29.

Louie, M. M. and Kolaczyk, E. D. (2004). "On the covariance properties of certain multiscale spatial processes." *Statistics and Probability Letters*, 66, 407–416.

— (2006a). "Multiscale detection of localized anomalous structure in aggregate disease incidence data." *Statistics in Medicine*, 25, 787–810.

— (2006b). "A multiscale method for disease mapping in spatial epidemiology." *Statistics in Medicine*, 25, 1287–1308.

Luettgen, M. R., Karl, W. C., and Willsky, A. S. (1994). "Efficient multiscale regularization with applications to the computation of optical flow." *IEEE Transactions on Image Processing*, 3, 41–63.

Luettgen, M. R., Karl, W. C., Willsky, A. S., and Tenney, R. R. (1993). "Multiscale Representations of Markov random fields." *IEEE Transactions on Signal Processing*, 41, 3377–3395.

Luettgen, M. R. and Willsky, A. S. (1995a). "Likelihood calculation for a class of multiscale stochastic models, with application to texture discrimination." *IEEE Transactions on Image Processing*, 4, 194–207.

— (1995b). "Multiscale smoothing error models." *IEEE Transactions on Automatic Control*, 40, 173–175.

Mallat, S. G. (1989). "Multiresolution approximations and the wavelet orthonormal bases of $L^2(R)$." *Transactions of the American Mathematical Society*, 315, 69–87.

— (1999). *A Wavelet Tour of Signal Processing*. 2nd ed. San Diego: Academic Press.

Mandelbrot, B. B. (1999). *Multifractals and 1/F Noise: Wild Self-affinity in Physics*. San Francisco: W. H. Freeman.

Marinari, E. and Parisi, G. (1992). "Simulated tempering: A new Monte Carlo scheme." *Europhysics Letters*, 19, 451–458.

Matérn, B. (1986). *Spatial Variation*. 2nd ed. New York: Springer-Verlag.

Matheron, G. (1963). "Principles of geostatistics." *Economic Geology*, 58, 1246–1266.

McLeod, A. I. (1994). "Diagnostic checking periodic autoregression models with application." *Journal of Time Series Analysis*, 15, 221–233.

Metropolis, N., Rosenbluth, A. W., Rosenbluth, M. N., Teller, A. H., and Teller, E. (1953). "Equation of state calculations by fast computing machines." *Journal of Chemical Physics*, 21, 1087–1092.

Morris, J. S., Vannucci, M., Brown, P. J., and Carroll, R. J. (2003). "Wavelet-based nonparametric modeling of hierarchical functions in colon carcinogenesis." *Journal of the American Statistical Association*, 98, 573–583.

Mugglin, A. S. and Carlin, B. P. (1998). "Hierarchical modeling in geographic information systems: Population interpolation over incompatible zones." *Journal of Agricultural, Biological, and Environmental Statistics*, 3, 111–130.

Mugglin, A. S., Carlin, B. P., and Gelfand, A. E. (2000). "Fully model-based approaches for spatially misaligned data." *Journal of the American Statistical Association*, 95, 877–887.

Müller, P. and Vidakovic, B., eds. (1999a). *Bayesian Inference in Wavelet Based Models*. Lecture Notes in Statistics. New York: Springer-Verlag.

Müller, P. and Vidakovic, B. (1999b). "MCMC methods in wavelet shrinkage: Non-equally spaced regression, density and spectral density estimation." In *Bayesian Inference in Wavelet Based Models*, eds. P. Müller and B. Vidakovic, 187–202. New York: Springer-Verlag.

Nason, G. P. (1998). "WaveThresh3 software." Technical Report, Department of Mathematics, University of Bristol.

Neal, R. (1997). "Monte Carlo implementation of Gaussian process models for Bayesian regression and classification." Technical Report 9702, Dept. of Computer Science, University of Toronto.

Neal, R. M. (1999). "Regression and classification using Gaussian process priors." In *Bayesian Statistics 6*, eds. J. M. Bernardo, J. O. Berger, A. P. Dawid, and A. F. M. Smith, 475–501. Oxford: Clarendon Press.

Neuman, S. P. and Yakowitz, S. (1979). "A statistical approach to the problem of aquifer hydrology: 1. Theory." *Water Resources Research*, 15, 845–860.

Neuwald, A. F. and Liu, J. S. (2004). "Gapped alignment of protein sequence motifs through Monte Carlo optimization of a hidden Markov model." *BMC Bioinformatics*, 5. Art. No. 157.

Nowak, R. D. (1999). "Multiscale hidden Markov models for Bayesian image analysis." In *Bayesian Inference in Wavelet Based Models*, eds. P. Müller and B. Vidakovic, 243–265. New York: Springer-Verlag.

Nowak, R. D. and Kolaczyk, E. D. (2000). "A statistical multiscale framework for Poisson inverse problems." *IEEE Transactions on Information Theory*, 46, 1811–1825.

O'Hagan, A. (1991). "Bayes-Hermite quadrature." *Journal of Statistical Planning and Inference*, 29, 145–260.

Oliver, D. S. (1994). "Incorporation of transient pressure data into reservoir characterization." *In Situ*, 18, 243–275.

Oliver, D. S., Cunha, L. B., and Reynolds, A. C. (1997). "Markov chain Monte Carlo methods for conditioning a permeability field to pressure data." *Mathematical Geology*, 29, 61–91.

Openshaw, S. and Taylor, P. (1979). "A million or so correlation coefficents." In *Statistical Methods in the Spatial Sciences*, ed. N. Wrigley, 127–133. London: Pion.

Paciorek, C. (2003). "Nonstationary Gaussian Processes for Regression and Spatial Modelling." Ph.D. thesis, Carnegie Mellon University, Department of Statistics.

Palm, F. C. and Nijman, T. E. (1984). "Missing observations in the dynamic regression model." *Econometrica*, 52, 1415–1435.

Park, C. G., Vannucci, M., and Hart, H. D. (2005). "Bayesian methods for wavelet series in single-index models." *Journal of Computation and Graphical Statistics*, 14, 1–25.

Pelletier, D. (2006). "Regime switching for dynamic correlations." *Journal of Econometrics*, 131, 445–473.

Peña, D., Tiao, G. C., and Tsay, R. (2000). *A Course in Time Series Analysis*. New York: John Wiley and Sons.

Pensky, M., Vidakovic, B., and De Canditiis, D. (2006). "Bayesian decision theoretic scale-adaptive estimation of spectral density." Technical Report, ISyE, Georgia Institute of Technology.

Pérez, P. and Heitz, F. (1996). "Restriction of a Markov random field on a graph and multiresolution statistical image modeling." *IEEE Transactions on Information Theory*, 42, 180–190.

Petris, G. and West, M. (1998). "Bayesian time series modelling with long-range dependence." Technical Report 686, Carnegie Mellon University.

Pizurica, A., Philips, W., Lemahieu, I., and Acheroy, M. (2002). "A joint inter and intrascale statistical model for Bayesian wavelet based image denoising." *IEEE Transactions on Image Processing*, 11, 545–557.

Priestley, M. B. (1981). *Spectral Analysis and Time Series*. San Diego: Academic Press.

Richardson, S. and Green, P. J. (1997). "On Bayesian analysis of mixtures with an unknown number of components (Disc: P758-792) (Corr: 1998V60 P661)." *Journal of the Royal Statistical Society, Series B*, 59, 731–758.

Ripley, B. D. (1981). *Spatial Statistics*. New York: John Wiley and Sons.

Robert, C. P. and Casella, G. (2005). *Monte Carlo Statistical Methods*. 2nd ed. New York: Springer-Verlag.

Roberts, G. O. and Gilks, W. R. (1994). "Convergence of adaptive direction sampling." *Journal of Multivariate Analysis*, 49, 287–298.

Rosenthal, J. S. (2000). "Parallel computing and Monte Carlo algorithms." *Far East Journal of Theoretical Statistics*, 4, 207–236.

Royle, J. A., Berliner, L. M., Wikle, C. K., and Milliff, R. (1999). "A hierarchical spatial model for constructing wind fields from scatterometer data in the Labrador Sea." In *Case Studies in Bayesian Statistics*, vol. IV, 367–382. New York: Springer-Verlag.

Rue, H. and Held, L. (2005). *Gaussian Markov Random Fields*. Boca Raton, FL: Chapman and Hall.

Sacks, J., Welch, W. J., Mitchell, T. J., and Wynn, H. P. (1989). "Design and analysis of computer experiments." *Statistical Science*, 4, 409–435.

Sampson, P. D. and Guttorp, P. (1992). "Nonparametric estimation of nonstationary spatial covariance structure." *Journal of the American Statistical Association*, 87, 108–119.

Santner, T. J., Williams, B. J., and Notz, W. I. (2003). *The Design and Analysis of Computer Experiments*. New York: Springer-Verlag.

Saquib, S. S., Bouman, C. A., and Sauer, K. (1996). "A non-homogeneous MRF model for multiresolution Bayesian estimation." In *IEEE International Conference on Image Processing*, vol. 2, 445–448. New York: IEEE.

Sarkar, U., Nascimento, S. F., Barbosa, R., Martins, R., Nuevo, H., Kalafanos, I., Grunstein, I., Flannery, B., Dias, J., Riley, L. W., Reis, M. G., and Ko, A. I. (2002). "A population-based case-control investigation of risk factors for leptospirosis during an urban epidemic." *American Journal of Tropical Medicine and Public Hygiene*, 66, 605–610.

Schmidt, A. M. and Gamerman, D. (1997). "Temporal aggregation in dynamic linear models." *Journal of Forecasting*, 16, 293–310.

Schmidt, A. M. and Gelfand, A. E. (2003). "A Bayesian coregionalization approach for multivariate pollutant data." *Journal of Geophysical Research–Atmospheres*, 108, 8783.

Schmidt, A. M. and O'Hagan, A. (2003). "Bayesian inference for nonstationary spatial covariance structure via spatial deformations." *Journal of the Royal Statistical Society, Series B*, 65, 743–758.

Scott, S. L. (1999). "Bayesian analysis of a two-state Markov modulated Poisson process." *Journal of Computational and Graphical Statistics*, 8, 662–670.

— (2002). "Bayesian methods for hidden Markov models: Recursive computing in the 21st century." *Journal of the American Statistical Association*, 97, 337–351.

— (2004). "A Bayesian paradigm for designing network intrusion systems." *Computational Statistics and Data Analysis*, 45, 69–83.

Scott, S. L., James, G. M., and Sugar, C. A. (2005). "Hidden Markov models for longitudinal comparisons." *Journal of the American Statistical Association*, 100, 359–369.

Sendur, L., Maxim, V., Whitcher, B., and Bullmore, E. (2005). "Multiple hypothesis mapping of functional MRI data in complex and orthogonal wavelet domains." *IEEE Transactions in Signal Processing*, 53, 3413–3426.

Shafer, G. (1981). "Jeffrey's rule of conditioning." *Philosophy of Science*, 48, 337–362.

Sirigos, J., Fakotakis, N., and Kokkinakis, G. (2002). "A hybrid syllable recognition system based on vowel spotting." *Speech Communication*, 38, 427–440.

Skyrms, B. (1980). *Causal Necessity*. New Haven, CT: Yale University Press.

Smith, R. L. (1993). "Long-range dependence and global warming." In *Statistics for the Environment*, eds. V. Barnett and F. Turkman, 141–161. Chichester: Wiley.

Sokal, A. D. (1987). "Monte Carlo methods in statistical mechanics: foundations and new algorithms." *Cours de Troisiéme Cycle de la Physique en Suisse Romande*. Lausanne.

Stein, M. L. (1999). *Interpolation of Spatial Data: Some Theory for Kriging*. New York: Springer-Verlag.

Stephen, K. D. and Dalrymple, M. (2002). "Reservoir simulations developed from an outcrop of incised valley fill strata." *AAPG Bulletin*, 86, 797–822.

Stroud, J. R., Müller, P., and Sansó, B. (2001). "Dynamic models for spatio-temporal data." *Journal of the Royal Statistical Society, Series B*, 63, 673–689.

Sun, D., Tsutakawa, R. K., and Speckman, P. L. (1999). "Posterior distribution of hierarchical models using CAR(1) distributions." *Biometrika*, 86, 341–350.

Tanner, M. A. (1993). *Tools for Statistical Inference: Methods for the Exploration of Posterior Distributions and Likelihood Functions*. 3rd ed. New York: Springer-Verlag.

Telser, L. G. (1967). "Discrete samples and moving sums in stationary stochastic processes." *Journal of the American Statistical Association*, 62, 484–499.

Thiébaux, H. J. and Pedder, M. A. (1987). *Spatial Objective Analysis with Applications in Atmospheric Science*. London: Academic Press.

Vannucci, M. (2007). *Wavelets in Statistics with Applications*. New York: Springer-Verlag. In press.

Vannucci, M., Brown, P. J., and Fearn, T. (2003). "A decision theoretical approach to wavelet regression on curves with a high number of regressors." *Journal of Statistical Planning and Inference*, 112, 195–212.

Vannucci, M. and Corradi, F. (1999a). "Covariance structure of wavelet coefficients: Theory and models in a Bayesian perspective." *Journal of the Royal Statistical Society, Series B*, 61, 971–986.

— (1999b). "Modeling dependence in the wavelet domain." In *Bayesian Inference in Wavelet Based Models*, eds. P. Müller and B. Vidakovic, 171–186. New York: Springer-Verlag.

Vannucci, M. and Liò, P. (2001). "Non-decimated wavelet analysis of biological sequences: Applications to protein structure and genomics." *Sankhyã*, 63, 218–233.

Vardi, Y., Shepp, L., and Kaufman, L. (1985). "A statistical model for positron emission tomography." *Journal of the American Statistical Association*, 80, 8–25.

Vasco, D. W. and Datta-Gupta, A. (1999). "Asymptotic solutions for solute transport: A formalism for tracer tomography." *Water Resources Research*, 35, 1–16.

Vasco, D. W., Yoon, S., and Datta-Gupta, A. (1998). "Integrating dynamic data into high-resolution reservoir models using streamline-based analytic sensitivity coefficients." Society of Petroleum Engineers 1998 Annual Technical Conference, SPE 49002.

Ver Hoef, J. M. and Barry, R. P. (1998). "Constructing and fitting models for cokriging and multivariable spatial prediction." *Journal of Statistical Planning and Inference*, 69, 275–294.

Vidakovic, B. (1998). "Nonlinear wavelet shrinkage with Bayes rules and Bayes factors." *Journal of the American Statistical Association*, 93, 173–179.

— (1999). *Statistical Modeling by Wavelets*. New York: Wiley.

Vidakovic, B. and Müller, P. (1995). "Wavelet shrinkage with affine Bayes rules with applications." Technical Report 95-34, Institute of Statistics and Decision Sciences, Duke University.

Viterbi, A. J. (1967). "Error bounds for convolutional codes and an asymptotically optimum decoding algorithm." *IEEE Transactions on Information Theory*, 13, 260–269.

Wackernagel, H. (1998). *Multivariate Geostatistics*. Berlin: Springer.

Wakefield, J. (2004). "A critique of statistical aspects of ecological studies in spatial epidemiology." *Environmental and Ecological Statistics*, 11, 31–54.

Weir, I. (1997). "Fully Bayesian reconstructions from single photon emission computed tomography." *Journal of the American Statistical Association*, 92, 49–60.

West, M. and Harrison, J. (1997). *Bayesian Forecasting and Dynamic Models*. 2nd ed. New York: Springer-Verlag.

Wikle, C. K. and Berliner, L. M. (2005). "Combining information across spatial scales." *Technometrics*, 47, 80–91.

Wikle, C. K., Berliner, L. M., and Cressie, N. (1998). "Hierarchical Bayesian space-time models." *Environmental and Ecological Statistics*, 5, 117–154.

Williams, C. K. I. and Rasmussen, C. E. (1996). "Gaussian processes for regression." In *Advances in Neural Information Precessing Systems 8*, eds. D. S. Tourestzky, M. C. Mozer, and M. E. Haeelmo. Cambridge, MA: MIT Press.

Willsky, A. S. (2002). "Multiresolution Markov models for signal and image processing." *Proceedings of the IEEE*, 90, 1396–1458.

Wong, W. H. (1995). "Comment (on Bayesian computation and stochastic systems)." *Statistical Science*, 10, 52–53.

Working, H. (1960). "Note on the correlation of first differences of averages in a random chain." *Econometrica*, 28, 916–918.

Wornell, G. W. (1990). "A Karhunen-Loéve-like expansion for 1/f processes via wavelets." *IEEE Transactions on Information Theory*, 36, 859–861.

Xue, G. and Datta-Gupta, A. (1996). "A new approach to seismic data integration using optimal non-parametric transformations." Society of Petroleum Engineers 1996 Annual Technical Conference, SPE 36500.

Yeh, W. W. (1986). "Review of parameter identification in groundwater hydrology: The inverse problem." *Water Resources Research*, 22, 95–108.

Yoon, S. (2000). "Dynamic data integration into high resolution reservoir models using streamline-based inversion." Ph.D. thesis, Texas A&M University, Department of Petroleum Engineering.

Yoon, S., Malallah, A. H., Datta-Gupta, A., Vasco, D. W., and Behrens, R. A. (1999). "A multiscale approach to production data integration using streamline models." Technical Report 56653, Society of Petroleum Engineers.

Index

Springer Series in Statistics   *(continued from p. ii)*

Bayesian Core: A Practical Approach to Computational Bayesian Statistics

Jean-Michel Marin and Christian P. Robert

This Bayesian modeling book is intended for practitioners and applied statisticians looking for a self-contained entry to computational Bayesian statistics. Focusing on standard statistical models and backed up by discussed real datasets available from the book website, it provides an operational methodology for conducting Bayesian inference, rather than focusing on its theoretical justifications. Special attention is paid to the derivation of prior distributions in each case and specific reference solutions are given for each of the models.

2007. 270 pp. (Springer Texts in Statistics) Hardcover
ISBN 978-0-387-38979-0

Finite Mixture and Markov Switching Models

Sylvia Frühwirth-Schnatter

The past decade has seen powerful new computational tools for modeling which combine a Bayesian approach with recent Monte simulation techniques based on Markov chains. This book is the first to offer a systematic presentation of the Bayesian perspective of finite mixture modeling. The book is designed to show finite mixture and Markov switching models are formulated, what structures they imply on the data, their potential uses, and how they are estimated.

2006. 492 pp. (Springer Series in Statistics) Hardcover
ISBN 978-0-387-32909-3

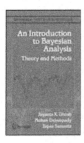

An Introduction to Bayesian Analysis

Jayanta K. Ghosh, Mohan Delampady, and Tapas Samanta

This is a graduate level textbook on Bayesian analysis blending modern Bayesian theory, methods, and applications. Starting from basic statistics, undergraduate calculus and linear algebra, ideas of both subjective and objective Bayesian analysis are developed to a level where real-life data can be analyzed using the current techniques of statistical computing. Advances in both low-dimensional and high-dimensional problems are covered, as well as important topics such as empirical Bayes and hierarchical Bayes methods and Markov chain Monte Carlo (MCMC) techniques.

2006. 365 pp. (Springer Texts in Statistics) Hardcover
ISBN 978-0-387-40084-6

Printed in the United States
By Bookmasters